Lecture Notes in Computer Science 6325

Commenced Publication in 1973
Founding and Former Series Editors:
Gerhard Goos, Juris Hartmanis, and Jan van Leeuwen

Uwe Aßmann   Andreas Bartho
Christian Wende (Eds.)

Reasoning Web

Semantic Technologies for Software Engineering

6th International Summer School 2010
Dresden, Germany, August 30 - September 3, 2010
Tutorial Lectures

 Springer

Volume Editors

Uwe Aßmann
Technische Universität Dresden, Fakultät Informatik
Institut für Software- und Multimediatechnik
Lehrstuhl Softwaretechnologie
01062 Dresden, Germany
E-mail: uwe.assmann@tu-dresden.de

Andreas Bartho
Technische Universität Dresden, Fakultät Informatik
Institut für Software- und Multimediatechnik
Lehrstuhl Softwaretechnologie
01062 Dresden, Germany
E-mail: andreas.bartho@tu-dresden.de

Christian Wende
Technische Universität Dresden, Fakultät Informatik
Institut für Software- und Multimediatechnik
Lehrstuhl Softwaretechnologie
01062 Dresden, Germany
E-mail: c.wende@tu-dresden.de

Library of Congress Control Number: 2010933248

CR Subject Classification (1998): H.4, H.3, I.2, H.5, C.2, D.2

LNCS Sublibrary: SL 3 – Information Systems and Application, incl. Internet/Web
and HCI

ISSN   0302-9743
ISBN-10   3-642-15542-1 Springer Berlin Heidelberg New York
ISBN-13   978-3-642-15542-0 Springer Berlin Heidelberg New York

springer.com

© Springer-Verlag Berlin Heidelberg 2010
Printed in Germany

Typesetting: Camera-ready by author, data conversion by Scientific Publishing Services, Chennai, India
Printed on acid-free paper   06/3180

Preface

Welcome to the proceedings of Reasoning Web 2010 which was held in Dresden.

Reasoning Web is a summer school series on theoretical foundations, contemporary approaches, and practical solutions for reasoning in a Web of Semantics. It has established itself as a meeting point for experts from research institutes and industry, as well as students undertaking their PhDs in related fields. This volume contains tutorial notes of the sixth school in the series, held from August 30 to September 3, 2010.

This year, the school focused on applications of semantic technologies in software engineering and the reasoning technologies appropriate for such an endeavor. As it turns out, semantic technologies in software engineering are not so easily applied, and several issues must be resolved before software modeling can benefit from reasoning. First, reasoning has to be fast and scalable, since models and programs can be quite large and voluminous. Since many reasoning languages are exponential or NP-complete, approximation, incrementalization, and other optimization techniques are extremely important. Second, software engineering needs to model software systems, in contrast to modeling domains of the world. Thus, the modeling techniques are prescriptive rather than descriptive [1], which influences the way models are reasoned about. When a software system is modeled, its behavior is *prescribed* by the model, that is, "the truth is in the model" [2]; when a domain of the world is described, its behavior cannot be *prescribed*, only *described* by the model ("the truth is in the world"). Therefore, reasoning has to distinguish between prescriptiveness and descriptiveness, leading to different assumptions about the closeness or openness of the world (closed-world assumption, CWA vs. open-world assumption, OWA). Third, while software modeling languages usually conform to a 4-level metalevel hierarchy (with objects on level M0, models on M1, metamodels on M2, metalanguages on M3), ontology languages usually only distinguish TBox and ABox. Different metalanguages are used, different repositories, and different strategies for integration with application code. Basically, ontology and software modeling worlds are two technological spaces [3], and these spaces have to be bridged on each of the M0-M3 levels. Since bridging often requires transformations of models to ontologies and ontologies to models, flexible glue technologies are looked for that hide the transformations from software developers.

These three requirements form only the top of the iceberg. There are many more problems below the surface, and the lecturers of the school attempt to provide answers for at least the following questions:

- How can we limit the complexity of reasoning to polynomial time? For this purpose, *expressive description logics* have been developed and transfered to OWL-EL, a profile for OWL ensuring polynomial complexity. Many of these contributions were achieved here in Dresden, and Anni-Yasmin Turhan from the Description Logic group presented them in the tutorial "Reasoning and Explanation in $\mathcal{EL}$ and in Expressive Description Logics."

– How can we reason with both rules and ontologies (hybrid reasoning)? Are we able to reason in an integrated fashion about domain models (which can easily be described by ontologies), requirements specifications (which talk about issues like business rules), and architecture specifications (which talk about rules for architectural styles)? What happens if a business rule from the requirement specification accesses the domain ontology? These fundamental questions of hybrid reasoning were taken up in the lecture on "Hybrid Reasoning with Non-Monotonic Rules" by Włodzimierz Drabent from the Polish Academy of Sciences and Linköping University.

– How can ontologies be embedded into the model-driven software development process (MDSD)? How can software models be checked on additional constraints? Can we model in an integrated fashion, that is, model in software languages while adding constraints from an ontology language? Which bridging technologies exist? These questions were discussed in the tutorial "Model-Driven Engineering with Ontology Technologies" by Steffen Staab, Tobias Walter, Gerd Gröner, and Fernando Silva Parreiras from the University of Koblenz-Landau.

– Can we use other techniques to speed up reasoning, such as approximation, incrementalization, or database optimization technology? How do we integrate requirements ontologies (which talk about the domains of the world with OWA) with system architecture specifications (which talk about systems with CWA)? Jeff Z. Pan from the University of Aberdeen discussed these issues in his talk "Scalable DL Reasoning" [4].

– Not only data definitions and their languages, but also query languages have to be bridged. Problems like OWA/CWA need to be taken into account while querying syntactic (model) data and semantic data. An example for a bridging technology for two leading query languages from the ontology and modeling worlds (SPARQL and GReQL) was presented in the tutorial "Bridging Query Languages in Semantic and Graph Technologies" by Hannes Schwarz and Jürgen Ebert from the University of Koblenz-Landau.

Apart from these more conceptual tutorials, the school featured a strong discussion of application areas in software engineering.

– Jens Lemcke, Tirdad Rahmani, and Andreas Friesen from SAP Research Karlsruhe reported on an application of ontologies in "Semantic Business Process Engineering." SAP has developed a reasoning-based refinement method for business process specifications, in which a concrete, executable workflow can be shown to be a behavioral refinement of an abstract business process. This method is very useful for enterprise process modeling, which is a focus area of SAP.

– Krzysztof Miksa, Pawel Sabina, and Marek Kasztelnik from Comarch (Krakow) showed how ontologies can be applied for the checking of constraints in domain-specific models in network device configuration ("Combining Ontologies with Domain Specific Languages: A Case Study from Network Configuration Software"). Comarch is one of the leading providers for telecommunication network software and needs to control huge domain models.

– Michael Schroeder from Technische Universität Dresden presented "Semantic Search Engines." Transinsight, his start-up company, has successfully built several semantic search machines for different biomedical applications, for instance, GoWeb [5].

– There were two smaller tutorials on "Semantic Service Engineering in the TEXO and Aletheia Projects" (SAP Dresden, Ralf Ackermann) and "Semantic Wikis" (Sören Auer, Leipzig University).

We believe that the application of reasoning technologies in software engineering will be a fruitful field in the future. The summer school clearly showed that a number of challenges exist, but also that they can be overcome with an appropriate bridging technology.

We would like to thank all lecturers of the Reasoning Web Summer School 2010 for their interesting and inspiring tutorials. We also thank the Program Committee members for executing their reviewing duties, in particular under the given time constraints in spring 2010. Many thanks go to Christiane Wagner and Gundula Gelfert from CMD Dresden for the organization of the school. Thanks also to Steffen Hölldobler and the summer school "Computational Logic," running in parallel with Reasoning Web this year [6], for their openness to share social activities. Finally, we also thank our funding agencies: this volume has been supported by the European Commission and by the Swiss Federal Office for Education and Science within the 7th Framework Programme project MOST number 216691.

September 2010   Uwe Aßmann
                                                                             Andreas Bartho
                                                                             Christan Wende

1. Ed Seidewitz, "What Models Mean", IEEE Software, September 2003
2. Jean-Marie Favre, Foundations of Model (Driven) (Reverse) Engineering: Models - Episode I, Stories of the Fidus Papyrus and of the Solarus. Dagstuhl Seminar 04101 on "Language Engineering for Model-Driven Software Development", Dagstuhl, Germany, February 29-March 5, 2004, Appeared in DROPS 04101, ISSN 1862-4405, IBFI. http://megaplanet.org/jean-marie-favre
3. Ivan Kurtev, Jean Bezivin, Mehmet Aksit, Technological Spaces: an Initial Appraisal, CoopIS, DOA'2002 Federated Conferences, Industrial track, Irvine, 2002
4. Yuan Ren, Jeff Z. Pan and Yuting Zhao. Soundness Preserving Approximation for TBox Reasoning. In Proc. of the 25th AAAI Conference Conference (AAAI2010). 2010. http://www.abdn.ac.uk/~csc280/pub/RPZ2010.pdf
5. Heiko Dietze and Michael Schroeder. GoWeb: a semantic search engine for the life science web. Biotechnology Center (BIOTEC), Technische Universität Dresden, 01062, Dresden, Germany. BMC Bioinformatics 2009, vol. 10(Suppl 10):S7, http://www.biomedcentral.com/1471-2105/10/S10/S7
6. Summer School Computational Logic 2010. http://www.computational-logic.org

Organization

Program Committee

José Júlio Alferes	University of Lisbon (Portugal)
Michael Altenhofen	SAP Research CEC Karlsruhe (Germany)
Uwe Aßmann	Technische Universität Dresden (Germany)
Colin Atkinson	University of Mannheim (Germany)
François Bry	Ludwig-Maximilians-Universität München (Germany)
Diego Calvanese	Free University of Bozen (Italy)
Carlos Damásio	University of Lisbon (Portugal)
Thomas Eiter	Vienna University of Technology (Austria)
Andreas Friesen	SAP Research CEC Karlsruhe (Germany)
Dragan Gašević	Athabasca University (Canada)
Jakob Henriksson	Intelligent Automation, Inc. (USA)
Carsten Lutz	University of Bremen (Germany)
Jan Małuszyński	University of Linköping (Sweden)
Daniel Oberle	SAP Research CEC Karlsruhe (Germany)
Jeff Z. Pan	University of Aberdeen (UK)
Renate Schmidt	University of Manchester (UK)
Michael Sintek	DFKI GmbH, Kaiserslautern (Germany)

Local Organization

Andreas Bartho	Technische Universität Dresden (Germany)
Christian Wende	Technische Universität Dresden (Germany)
Christiane Wagner	Technische Universität Dresden (Germany)
Gundula Gelfert	CMD Congress Management GmbH, Dresden (Germany)

Proceedings Chair

Uwe Aßmann	Technische Universität Dresden (Germany)

Sponsoring Institutions

Table of Contents

Reasoning and Explanation in $\mathcal{EL}$ and in Expressive Description Logics

Anni-Yasmin Turhan

Theoretical Computer Science,
TU Dresden, Germany
turhan@tcs.inf.tu-dresden.de

Abstract. Description Logics (DLs) are the formalism underlying the standard web ontology language OWL 2. DLs have formal semantics which are the basis for powerful reasoning services. In this paper, we introduce the basic notions of DLs and the techniques that realize subsumption—the fundamental reasoning service of DL systems. We discuss two reasoning methods for this service: the tableau method for expressive DLs such as $\mathcal{ALC}$ and the completion method for the light-weight DL $\mathcal{EL}$. We also present methods for generating explanations for computed subsumption relationships in these two DLs.

1 Introduction

The ontology language for the semantic web OWL provides means to describe entities of a application domain in an ontology. The underlying formalism for OWL are Description Logics, which have well-defined syntax and formal semantics. The recent version of the W3C standard OWL 2.0 has four language variants: the OWL 2 language itself and three profiles. The latter are light-weight ontology languages of relatively low expressivity and that are tailored to be efficient for specific reasoning tasks. We are interested in the reasoning task of computing subsumption, i.e., sub- and super-class relationships, and providing explanations for the obtained reasoning results. In this paper, we discuss reasoning techniques for computing subsumption relationships for the core description logics underlying the OWL 2 language: $\mathcal{ALC}$ and the core description logics underlying the EL profile: $\mathcal{EL}$. The EL profile is particularly suitable for applications with ontologies that define very large numbers of classes and that need subsumption as the main inference service. Based on the reasoning techniques for subsumption, we discuss methods to compute explanations for detected subsumption relationships in $\mathcal{ALC}$ and $\mathcal{EL}$. Before we turn to the reasoning techniques, we give general overview of Description Logics.

Description Logics (DLs) [6] are a family of knowledge representation formalisms that have formal semantics. This family of logics is tailored towards representing terminological knowledge of an application domain in a structured and formally well-understood way. Description logics allow users to define important notions, such as classes or relations of their application domain in terms of concepts and roles. These concepts (unary predicates) and roles (binary predicates) then restrict the way these classes and relations are interpreted. Based on these definitions, implicitly captured

U. Aßmann, A. Bartho, and C. Wende (Eds.): Reasoning Web 2010, LNCS 6325, pp. 1–27, 2010.

knowledge can be inferred from the given descriptions of concepts and roles, as for instance sub-class or instance relationships.

The name *Description Logics* is motivated by the fact that classes and relations are defined in terms of concept *descriptions*. These concept descriptions are complex expressions built from atomic concepts and atomic roles using the concept constructors offered by the particular DL in use. Based on their formal semantics, a whole collection of inference services has been defined and investigated for different DLs. DLs have been employed in various domains, such as databases, biomedical or context-aware applications [3, 96]. Their most notable success so far is probably the adoption of the DL-based language OWL[1] as standard ontology language for the Semantic Web [53].

Historically, DLs stem from knowledge representation systems such as *semantic networks* [85, 94] or *frame systems* [73]. These early knowledge representation systems were motivated by linguistic applications and allow to specify information from the domain of discourse. They offer methods to compute inheritance relations between the specified notions. Early frame-based systems and semantic networks both have operational semantics, i.e., the semantics of reasoning is given by its implementation. As a consequence, the result of the reasoning process depends on the implementation of the reasoner and thus the result may differ from system to system for the same input [95]. To remedy this, DLs and their reasoning services are based on formal semantics. The information about the application domain is represented in a declarative and unambiguous way. More importantly, the formal semantics of the reasoning services ensure predictable and thus reliable behavior of the DL reasoning systems—independent of the implementation.

The investigation of algorithms for reasoning services and their complexity is the main focus of the DL research community. Typically, one can distinguish the following phases of DL research during the last decades. In the late eighties, reasoning algorithms have been devised for DL systems that mostly were sound, but incomplete, i.e., they would return correct answers, but would not find *all* correct answers. This development was led by the belief that terminological reasoning is inherently intractable [79,80], and thus completeness was traded for tractability. These algorithms have been implemented in systems such as Classic [23,22,84] and Back [79,81]. During the nineties, sound and complete reasoning methods were investigated for the core inferences of DL systems: consistency and subsumption. *Consistency* assures that the specification of the concepts, roles and individuals are free of contradictions. For *subsumption* one computes super- and sub-concept relations from the given specifications of concepts and roles. The use of incomplete algorithms for these inferences has largely been abandoned in the DL community since then, mainly because of the problem that the behavior of the systems is no longer determined by the semantics of the description language: an incomplete algorithm may claim that a subsumption relationship does not hold, although it should hold according to the semantics.

The underlying technique for computing the basic DL inferences is the tableau method [37], which was adapted to DLs in [91]. This method was extended to more and more expressive DLs (for an overview, see [17]). The gain in expressiveness came at the cost

---

[1] http://www.w3.org/TR/owl-features/

of higher complexity for the reasoning procedures—reasoning for the DLs investigated is PSpace-complete or even ExpTime-complete [66, 54, 98] (for an overview see [17, 31]).

Despite the high complexity, highly optimized DL reasoning systems were implemented based on the tableau method—most prominently the FACT system [49] and RACER [43]. These systems employed optimization methods developed for DL reasoning based on tableaux [7,48,58,45] and demonstrated that the high worst case complexities would hardly be encountered in practice [49, 52, 58, 42, 50, 100]. In fact, it turned out that these highly optimized implementations of the reasoning methods do perform surprisingly well on DL knowledge bases from practical applications.

Encouraged by these findings and driven by application needs researchers investigated tableau algorithms for even more expressive DLs [55,56,51,57] in the last decade. At the same time, the idea of the Semantic Web emerged and DLs became the basis for the W3C standardized web ontology language OWL [53,44]. This brought DLs into the attention of new users from various application areas, which in turn necessitated automated support of ontology services and motivated research on various new inferences for DLs. For instance,

- the generation of *explanations* of consequences that the DL reasoner detected [90, 83, 63, 61, 15],
- support for building ontologies by computing *generalizations* [10, 27, 18, 101, 35],
- *conjunctive queries* as a means to access the instance data of an ontology [76, 29, 30, 39, 82, 36, 67], and
- computing *modularizations* of an ontology as means to facilitate their reuse [38, 69, 33, 32, 70].

All of them are currently investigated reasoning services for DLs and most of them are implemented in specialized reasoners. At the same time, the need for faster reasoners for the afore mentioned basic inferences for DLs led to two developments. On the one hand, the new tableau-based reasoners for expressive DL were developed such as PELLET [93], FACT++ [99, 100] and RACERPRO [86] and new reasoning methods for expressive DLs were investigated and implemented such as resolution [74, 76] in KAON2 and hyper-tableau [77, 78] in HERMIT. On the other hand, *light-weight DLs*, which are DLs with relatively limited expressivity, but good computational properties for specific reasoning tasks were designed [13]. Reasoning even for large ontologies written in these DLs can be done efficiently, since the respective reasoning methods are tractable. There are two "families" of lightweight DLs: the $\mathcal{EL}$ family [25, 4, 5], for which the subsumption and the instance problem are polynomial, and the DL Lite family [28, 30], for which the instance problem and query answering are polynomial. A member of each of these families is the DL corresponding to one of the profiles of the OWL 2 standard.

In this paper, we examine the basic reasoning services for DLs for the light-weight DL $\mathcal{EL}$ and for expressive DLs. In the next section, we give basic definitions for the fundamental DLs $\mathcal{ALC}$ and $\mathcal{EL}$. We introduce basic notions such as concept descriptions, TBoxes and ABoxes and their semantics. Based on this, we define the central reasoning services common to most DL systems. In Section 3, we discuss the reasoning methods for basic reasoning problems: we describe the tableau method for $\mathcal{ALC}$ and the completion-based approach for $\mathcal{EL}$. In Section 4, we turn to another reasoning service, namely the computation of explanations for (probably unexpected) reasoning results. Again, we consider methods for expressive DLs and for $\mathcal{EL}$ for this task.

2 Basic Definitions

The central notion for DLs are *concept descriptions*, which can be built from concept names and so-called *concept constructors*. For instance, one can describe a course as an event given by a lecturer in the following way by a concept description:

$$\text{Event} \sqcap \exists \text{ given-by.Lecturer} \sqcap \exists \text{ has-topic.} \top$$

This concept description is a conjunction (indicated by $\sqcap$) of the concept Event, the existential restriction $\exists$ given-by.Lecturer and the existential restriction $\exists$ has-topic.$\top$. The first existential restriction consists of the role name given-by and concept Lecturer, which relates the Lecturer to the course. The latter existential restriction states that there is a topic (which is not specified).

In general, concept descriptions are built from the set of concept names N_C and the set of role names N_R using concept constructors. Every DL offers a different set of concept constructors. The DL $\mathcal{EL}$ allows only for the concept constructors that were used in the example concept description above.

Definition 1 (*$\mathcal{EL}$-concept descriptions*). *Let N_C be a set of concept names and N_R a set of role names. The set of $\mathcal{EL}$-concept descriptions is the smallest set such that*

- *all concept names are $\mathcal{EL}$-concept descriptions;*
- *if C and D are $\mathcal{EL}$-concept descriptions, then $C \sqcap D$ is also an $\mathcal{EL}$-concept description;*
- *if C is an $\mathcal{EL}$-concept description and $r \in N_R$, then $\exists r.C$ is also an $\mathcal{EL}$-concept description.*

If this set of concept constructors is extended to all Boolean connectors, i.e., extended by disjunction ($\sqcup$) and full negation ($\neg$), one obtains the DL $\mathcal{ALC}$. We can define $\mathcal{ALC}$-concept descriptions inductively.

Definition 2 (*$\mathcal{ALC}$-concept descriptions*). *Let N_C be a set of concept names and N_R a set of role names. The set of $\mathcal{ALC}$-concept descriptions is the smallest set such that*

- *all concept names are $\mathcal{ALC}$-concept descriptions;*
- *if C and D are $\mathcal{ALC}$-concept descriptions, then $\neg C$, $C \sqcap D$ and $C \sqcup D$ are also $\mathcal{ALC}$-concept descriptions;*
- *if C is an $\mathcal{ALC}$-concept description and $r \in N_R$, then $\exists r.C$ and $\forall r.C$ are also $\mathcal{ALC}$-concept descriptions.*

We call concept descriptions of the form $\exists r.C$ existential restrictions and concept descriptions of the form $\forall r.C$ value restrictions. The semantics of DL concept descriptions is given by means of interpretations.

Definition 3 (Semantics of $\mathcal{ALC}$-concept descriptions). *Let C and D be $\mathcal{ALC}$-concept descriptions and r a role name. An interpretation is a pair $\mathcal{I} = (\Delta^{\mathcal{I}}, \cdot^{\mathcal{I}})$ where the domain $\Delta^{\mathcal{I}}$ is a non-empty set and $\cdot^{\mathcal{I}}$ is a function that assigns to every concept name A a set $A^{\mathcal{I}} \subseteq \Delta^{\mathcal{I}}$ and to every role name r a binary relation $r^{\mathcal{I}} \subseteq \Delta^{\mathcal{I}} \times \Delta^{\mathcal{I}}$. This function is extended to complex $\mathcal{ALC}$-concept descriptions as follows:*

- $(C \sqcap D)^{\mathcal{I}} = C^{\mathcal{I}} \cap D^{\mathcal{I}}$;
- $(C \sqcup D)^{\mathcal{I}} = C^{\mathcal{I}} \cup D^{\mathcal{I}}$;
- $(\neg C)^{\mathcal{I}} = \Delta^{\mathcal{I}} \setminus C^{\mathcal{I}}$;
- $(\exists r.C)^{\mathcal{I}} = \{x \in \Delta^{\mathcal{I}} \mid \text{there is a } y \in \Delta^{\mathcal{I}} \text{ with } (x,y) \in r^{\mathcal{I}} \text{ and } y \in C^{\mathcal{I}}\}$; and
- $(\forall r.C)^{\mathcal{I}} = \{x \in \Delta^{\mathcal{I}} \mid \text{for all } y \in \Delta^{\mathcal{I}}, (x,y) \in r^{\mathcal{I}} \text{ implies } y \in C^{\mathcal{I}}\}$.

This definitions clearly also captures the semantics of the less expressive DL $\mathcal{EL}$. Both, $\mathcal{EL}$ and $\mathcal{ALC}$ also offer the *top-concept* $\top$, which is always interpreted as the whole domain $\Delta^{\mathcal{I}}$. In addition $\mathcal{ALC}$ also offers the *bottom concept* $\bot$, which is always interpreted as the empty set. Now, with the $\mathcal{ALC}$-concept constructors at hand, one can, for instance, characterize a graduate CS student by the following concept description:

$$\exists \text{ studies-subject. CS} \sqcap (\text{Master-Student} \sqcup \text{PhD-Student})$$

Concept description like these are the main building blocks to model terminological knowledge.

2.1 Terminological Knowledge

A name can be assigned to a concept description by a *concept definition*. For instance, we can write Course $\equiv$ Event $\sqcap$ $\exists$ given-by.Lecturer $\sqcap$ $\exists$ has-topic. $\top$ to supply a concept definition for the concept Course.

Definition 4 (Concept definition, general concept inclusion). *Let A be a concept name and C, D be (possibly) complex concept description.*

- *A concept definition is a statement of the form $A \equiv C$.*
- *A general concept inclusion (GCI for short) is a statement of the form $C \sqsubseteq D$.*

It is easy to see that every concept definition $A \equiv C$ can be expressed by two GCIs: $A \sqsubseteq C$ and $C \sqsubseteq A$. The terminological information expressed by GCIs is collected in the so-called TBox.

Definition 5 (TBox). *A finite set of GCIs is called a TBox. An interpretation is a model of a TBox $\mathcal{T}$, if it satisfies all GCIs, i.e., if $C^{\mathcal{I}} \subseteq D^{\mathcal{I}}$ for all $C \sqsubseteq D$ in $\mathcal{T}$.*

If all concept descriptions in a TBox $\mathcal{T}$ are from a description logic $\mathcal{L}$, then we call $\mathcal{T}$ a $\mathcal{L}$-TBox.

   If a concept definition $A \equiv C$ in a TBox uses a concept name B directly, i.e., B appears in C, or if B is used indirectly by the definitions of the names appearing in C, we say that the TBox is *cyclic*. Otherwise a TBox is *acyclic*.

Definition 6 (Unfoldable TBox). *A TBox $\mathcal{T}$ is a finite set of concept definitions that is acyclic and such that every concept name appears at most once on the left-hand side of the concept definitions in $\mathcal{T}$. Given a TBox $\mathcal{T}$, we call the concept name A a* defined concept, *if A occurs on the left-hand side of a concept definition in $\mathcal{T}$. All other concepts are called* primitive concepts.

One of the basic reasoning services in DL systems is to test for the *satisfiability* of a concept or a TBox, i.e., to test whether the information specified in it contains logical contradictions or not. In case the TBox contains a contradiction, any consequence can follow logically from the TBox. Moreover, if a TBox is not satisfiable, the specified information can hardly capture the intended meaning from an application domain. To test for satisfiability is often a first step for a user to check whether a TBox models something "meaningful".

Definition 7 (Concept satisfiability, TBox satisfiability). *Let C be a concept description and $\mathcal{T}$ a TBox. The concept description C is* satisfiable *iff it has a model, i.e., iff there exists an interpretation $\mathcal{I}$ such that $C^{\mathcal{I}} \neq \emptyset$. A TBox $\mathcal{T}$ is* satisfiable *iff it has a model, i.e., an interpretation that satisfies all GCIs in $\mathcal{T}$.*

If a concept or TBox is not satisfiable, it is called *unsatisfiable*. Other typical reasoning services offered in DL systems test for equivalence or inclusion relations between concepts. In the latter case, if one concept of the TBox models a more general category than another one, we say that this concept *subsumes* the other one.

Definition 8 (Concept subsumption, concept equivalence). *Let C, D be two concept descriptions and $\mathcal{T}$ a (possibly empty) TBox. The concept description C is* subsumed *by the concept description D w.r.t. $\mathcal{T}$ ($C \sqsubseteq_{\mathcal{T}} D$), iff $C^{\mathcal{I}} \subseteq D^{\mathcal{I}}$ holds in every model $\mathcal{I}$ of $\mathcal{T}$. Two concepts C, D are* equivalent w.r.t. $\mathcal{T}$ *($C \equiv_{\mathcal{T}} D$), iff $C^{\mathcal{I}} = D^{\mathcal{I}}$ holds for every model $\mathcal{I}$ of $\mathcal{T}$.*

The computation of the subsumption relations for all named concepts mentioned in the TBox $\mathcal{T}$ is called *classification* of the TBox $\mathcal{T}$ and yields the *concept hierarchy* of the TBox $\mathcal{T}$.

2.2 Assertional Knowledge

Facts about individuals from the application domain can be stated by *assertions*. There are two basic kinds of assertions for DL systems—one expresses that an individual belongs to a concept and the other one specifies that two individuals are related via a role. The set N_I is the set of all individual names.

Definition 9 (Assertion, ABox). *Let C be a concept description, $r \in NR$ a role name and i, j ($\{i, j\} \subseteq N_I$) be two individual names, then*

- *$C(i)$ is called a* concept assertion *and*
- *$r(i, j)$ is called a* role assertion.

An ABox $\mathcal{A}$ is a finite set of concept assertions and role assertions.

For instance, we can express that Dresden is a city located at the river Elbe by the following ABox:

$$\{ \text{City(Dresden), River(Elbe), located-at(Dresden, Elbe)} \}$$

If all concept descriptions in an ABox $\mathcal{A}$ are from a Description Logic $\mathcal{L}$, then we call $\mathcal{A}$ a $\mathcal{L}$-ABox. In order to capture ABoxes, the interpretation function is now extended to individual names. Each individual name is mapped by the interpretation function to an element of the domain $\Delta^{\mathcal{I}}$.

Definition 10 (Semantics of assertions, semantics of ABoxes). *Let C be a concept description, r a role name and i, j two individual names, then an interpretation $\mathcal{I}$ satisfies*

- *the concept assertion $C(i)$ if $i^{\mathcal{I}} \in C^{\mathcal{I}}$ and*
- *the role assertion $r(i, j)$ if $(i^{\mathcal{I}}, j^{\mathcal{I}}) \in r^{\mathcal{I}}$.*

An interpretation $\mathcal{I}$ is a model of an ABox $\mathcal{A}$, if $\mathcal{I}$ satisfies every assertion in $\mathcal{A}$.

A DL *knowledge base* $\mathcal{K}$ consists of an ABox $\mathcal{A}$ and a TBox $\mathcal{T}$. We write $\mathcal{K} = (\mathcal{T}, \mathcal{A})$. We can now test for the absence of contradictions in ABoxes.

Definition 11 (ABox consistency, instance of). *An ABox $\mathcal{A}$ is consistent w.r.t. a TBox $\mathcal{T}$, iff it has a model that is also a model for $\mathcal{T}$. The individual i is an instance of the concept description C w.r.t. an ABox $\mathcal{A}$ and a TBox $\mathcal{T}$ (we write $\mathcal{A} \models_{\mathcal{T}} C(i)$), iff $i^{\mathcal{I}} \in C^{\mathcal{I}}$ for all models $\mathcal{I}$ of $\mathcal{T}$ and $\mathcal{A}$.*

ABox realization is a reasoning service that computes for each individual i of an ABox $\mathcal{A}$ and a TBox $\mathcal{T}$ the set of all named concepts A appearing in $\mathcal{A}$ and $\mathcal{T}$ that (1) have i as an instance ($\mathcal{A} \models_{\mathcal{T}} A(i)$) and (2) that is least w.r.t. $\sqsubseteq_{\mathcal{T}}$.

Typically, all the reasoning services introduced in this section are implemented in DL systems. In Section 3, we discuss the reasoning algorithms for these inferences for $\mathcal{ALC}$ and in more detail for $\mathcal{EL}$. Before we do so, we survey some extensions of these two basic DLs.

2.3   Extensions of Basic DLs

The basic DL $\mathcal{ALC}$ has been extended in many ways and, as mentioned in the introduction, reasoning algorithms have been devised for many of these extensions, see [31]. We consider here now some of those extensions that are captured in the OWL 2 standard [102] and that are also covered in the OWL 2 EL profile [75]. The DLs underlying these standardized ontology languages are $\mathcal{SROIQ}$ [51] and $\mathcal{EL}^{++}$ [5], respectively. Both DLs allow to specify more information on roles.

A role r can be declared to be a *transitive role* in the TBox. The semantics is straight-forward. An interpretation $\mathcal{I}$ satisfies a transitive role declaration transitive(r) if $\{(a, b), (b, c)\} \subseteq r^{\mathcal{I}}$ implies $(a, c) \in r^{\mathcal{I}}$. Transitive roles can be used in concept descriptions. Assume that the role has-part is transitive, then the two axioms:

$$\text{Summer-school} \equiv \exists \, \text{has-part. Course}$$
$$\text{Course} \equiv \exists \, \text{has-part. Lesson}$$

imply that a Summer school has a part that is a lesson. The declaration of an *inverse role* applies to a role name r and yields its inverse r^{-1}, where the semantics is the obvious one, i.e.,

$$(r^{-1})^{\mathcal{I}} := \{(e, d) \mid (d, e) \in r^{\mathcal{I}}\}.$$

Using the inverse of the role attends, we can define the concept of a speaker giving a boring talk as

$$\text{Speaker} \sqcap \exists \text{gives}.(\text{Talk} \sqcap \forall \text{attends}^{-1}.(\text{Bored} \sqcup \text{Sleeping})).$$

Furthermore, it can be specified that a role is a super-role of another role by a *role inclusion axiom*. The set of all role inclusions form the *role hierarchy*. An interpretation $\mathcal{I}$ satisfies a role inclusion axiom $r \sqsubseteq s$ if $r^{\mathcal{I}} \subseteq s^{\mathcal{I}}$.

For instance, we might capture the fact that everybody who is attending something (a course) is also interested in this (course) by a role inclusion axiom

$$\text{attends} \sqsubseteq \text{interested-in}.$$

DL researchers have introduced many additional constructors to the basic DL $\mathcal{ALC}$ and investigated various DLs obtained by combining such constructors. Here, we only introduce qualified number restrictions as example for additional concept constructors. This extension is covered also in the DL $\mathcal{SROIQ}$, but not in $\mathcal{EL}^{++}$. See [1] for an extensive list of additional concept and role constructors.

Qualified number restrictions are of the form $(\geq n\, r.C)$ (at-least restriction) and $(\leq n\, r.C)$ (at-most restriction), where $n \geq 0$ is a non-negative integer, $r \in N_R$ is a role name, and C is a concept description. The semantics of these additional constructors is defined as follows:

$$(\geq n\, r.C)^{\mathcal{I}} := \{d \in \Delta^{\mathcal{I}} \mid card(\{e \mid (d,e) \in r^{\mathcal{I}} \wedge e \in C^{\mathcal{I}}\}) \geq n\},$$
$$(\leq n\, r.C)^{\mathcal{I}} := \{d \in \Delta^{\mathcal{I}} \mid card(\{e \mid (d,e) \in r^{\mathcal{I}} \wedge e \in C^{\mathcal{I}}\}) \leq n\},$$

where $card(X)$ yields the cardinality of the set X. Using qualified number restrictions, we can define the concept of all persons that attend at most 20 talks, of which at least 3 have the topic DL:

$$\text{Person} \sqcap (\leq 20 \text{ attends}.\text{Talk}) \sqcap (\geq 3 \text{ attends}.(\text{Talk} \sqcap \exists \text{topic}.\text{DL})).$$

2.4 Relations of DLs to Other Logics

Description logics are logic-based knowledge representation formalisms. A natural question is how they are related to other logics. In fact, it is easy to see, given their semantics, that most description logics are a fragment of first order logic (FOL). Concept descriptions can be translated into FOL formulae with one free variable. Concept names can be interpreted as unary predicates and role names as binary relations, see for example [88, 68, 59]. An arbitrary $\mathcal{ALC}$-concept description can be translated into a FOL formula τ_x, where x is a free variable in the following way:

- $\tau_x(A) := A(x)$ for a concept name A,
- $\tau_x(\neg C) := \neg \tau_x(C)$,
- $\tau_x(C \sqcap D) := \tau_x(C) \wedge \tau_x(D)$,
- $\tau_x(C \sqcup D) := \tau_x(C) \vee \tau_x(D)$,
- $\tau_x(\exists r.C) := \exists y.(r(x,y) \vee \tau_y(C))$, where y is a variable different from x, and
- $\tau_x(\forall r.C) := \forall y.(r(x,y) \rightarrow \tau_y(C))$, where y is a variable different from x.

The intuition of the translation to FOL is that the formula $\tau_x(C)$ describes all domain elements d from $\Delta^{\mathcal{I}}$ that make the formula τ_x true if x is replaced by d. This clearly coincides with the interpretation of the concept description $C^{\mathcal{I}}$. The translation does not yield arbitrary FOL formulae, but formulae from the two-variable fragment [41] and the guarded fragment [40]. Both of which are known to be decidable.

Description Logics are closely related to modal logics (see e.g. [37, 21]). For instance, the DL $\mathcal{ALC}$ is a syntactic variant of the multimodal logic K, see [89]. The multimodal logic K introduces several box and diamond operators that are indexed with the name of the corresponding transition relation, which can be directly translated into $\mathcal{ALC}$ using role names corresponding to the transition relations.

Any $\mathcal{ALC}$ interpretation $\mathcal{I}$ can be viewed as a Kripke structure $K_{\mathcal{I}}$. The elements of the domain $w \in \Delta^{\mathcal{I}}$ correspond to possible worlds in $K_{\mathcal{I}}$. A propositional variable A is true in world w, iff $w \in A^{\mathcal{I}}$. There is a transition relation r in the Kripke structure from world w_1 to world w_2 iff $(w_1, w_2) \in r^{\mathcal{I}}$. Many theoretical results on reasoning in modal logics carry directly over to standard inferences in DLs due to this direct translation.

3 DL Reasoning

In this section we present reasoning methods for the DL reasoning problems defined in the last section: satisfiability and subsumption. These problems are decision problems and we devise decision procedures for them. Before we do so, we recall some general requirements that we would like to hold for such decision procedures. Such a procedure must be:

- *sound*, i.e., the positive answers should be correct;
- *complete*, i.e., the negative answers should be correct; and
- *terminating*, i.e., it should always give an answer in finite time.

Together these properties ensure that we always obtain an answer and that every given answer of the procedure is correct. These properties guarantee that applications built on top of these procedures are predictable and reliable. To employ the decision procedures in real world applications, we also would like our decision procedure to be

- *efficient*, i.e., it should be optimal w.r.t. the (worst-case) complexity of the problem, and
- *practical*, i.e., easy to implement and optimize, and behave well for application cases.

DL research has mostly been dedicated to design decision procedures that fulfill these requirements. The underlying techniques to realize reasoning procedures that we are considering in the following are the tableaux method for expressive DLs and completion for $\mathcal{EL}$.

3.1 Reasoning in Expressive DLs

By expressive DLs we refer to DLs that offer at least all Boolean constructors and that are thus closed under negation. For this kind of DLs, it is not necessary to design and implement different algorithms for the different reasoning problems introduced in the last section, since there exist polynomial time reductions, which only require the availability of the concept constructors conjunction and negation in the description language. For the TBox reasoning problems there are the following reductions:

- Subsumption can be be reduced in polynomial time to equivalence:

$$C \sqsubseteq_T D \text{ iff } C \sqcap D \equiv_T C.$$

- Equivalence can be be reduced in polynomial time to subsumption:

$$C \equiv_T D \text{ iff } C \sqsubseteq_T D \text{ and } D \sqsubseteq_T C.$$

- Subsumption can be be reduced in polynomial time to (un)satisfiability:

$$C \sqsubseteq_T D \text{ iff } C \sqcap \neg D \text{ is unsatisfiable w.r.t. } T.$$

- Satisfiability can be be reduced in polynomial time to (non-)subsumption:

$$C \text{ is satisfiable w.r.t. } T \text{ iff not } C \sqsubseteq_T \bot.$$

For reasoning problems w.r.t. ABoxes (and TBoxes) there are similar polynomial time reductions:

- Satisfiability can be be reduced in polynomial time to consistency:

$$C \text{ is satisfiable w.r.t. } T \text{ iff the ABox } \{C(a)\} \text{ is consistent w.r.t. } T.$$

- The instance problem can be reduced in polynomial time to (in)consistency:

$$A \models_T C(a) \text{ iff } A \cup \{\neg C(a)\} \text{ is inconsistent w.r.t. } T.$$

- Consistency can be reduced in polynomial time to the (non-)instance problem:

$$A \text{ is consistent w.r.t. } T \text{ iff } A \not\models_T \bot(a).$$

With these reductions at hand, it suffices to investigate a reasoning procedure for one of the reasoning problems. In this section, we restrict ourselves to unfoldable TBoxes, i.e., TBoxes without GCIs and cyclic definitions. We present a tableau algorithm for deciding ABox consistency in this setting. Such a tableau-based algorithm tries to construct a model for the ABox by breaking down the concept descriptions in the knowledge base and inferring new constraints on the elements of this model. The algorithm either stops because all attempts to build a model failed due to obvious contradictions, or it stops with a "canonical" model.

In a first step of the consistency test, negation is treated by transforming the concept description from the knowledge base into *negation normal form (NNF)*. This normal form pushes all negations into the description until they occur only in front of concept names, using de Morgan' rules.

The $\rightarrow_{\sqcap}$-rule
Condition: $\mathcal{A}$ contains $(C_1 \sqcap C_2)(x)$, but not both $C_1(x)$ and $C_2(x)$.
Action: $\mathcal{A}' := \mathcal{A} \cup \{C_1(x), C_2(x)\}$.

The $\rightarrow_{\sqcup}$-rule
Condition: $\mathcal{A}$ contains $(C_1 \sqcup C_2)(x)$, but neither $C_1(x)$ nor $C_2(x)$.
Action: $\mathcal{A}' := \mathcal{A} \cup \{C_1(x)\}$, $\mathcal{A}'' := \mathcal{A} \cup \{C_2(x)\}$.

The $\rightarrow_{\exists}$-rule
Condition: $\mathcal{A}$ contains $(\exists r.C)(x)$, but there is no individual name z such that $C(z)$
                   and $r(x,z)$ are in $\mathcal{A}$.
Action: $\mathcal{A}' := \mathcal{A} \cup \{C(y), r(x,y)\}$ where y is an individual name not occurring in $\mathcal{A}$.

The $\rightarrow_{\forall}$-rule
Condition: $\mathcal{A}$ contains $(\forall r.C)(x)$ and $r(x,y)$, but it does not contain $C(y)$.
Action: $\mathcal{A}' := \mathcal{A} \cup \{C(y)\}$.

Fig. 1. Tableau rules of the consistency algorithm for $\mathcal{ALC}$

Definition 12 ($\mathcal{ALC}$-negation normal form). *An $\mathcal{ALC}$-concept description is in $\mathcal{ALC}$-negation normal form (NNF) if the following rules have been applied exhaustively:*

$$\neg\bot \rightarrow \top \qquad\qquad \neg(C \sqcap D) \rightarrow (\neg C \sqcup \neg D) \qquad\qquad \neg(\exists r.C) \rightarrow (\forall r.\neg C)$$
$$\neg\top \rightarrow \bot \qquad\qquad \neg(C \sqcup D) \rightarrow (\neg C \sqcap \neg D) \qquad\qquad \neg(\forall r.C) \rightarrow (\exists r.\neg C)$$
$$\neg\neg C \rightarrow C$$

A TBox or an ABox is in NNF, if all concept descriptions appearing in it are in NNF.

The *size* of an $\mathcal{ALC}$-concept description is the number of occurrences of all concept and role names that appear in the concept description. The size of a TBox is the sum of the sizes of all the concept descriptions appearing in the TBox. Similarly, the size of an ABox is the sum of all the concept descriptions appearing the concept assertions plus the number of role assertions. Transforming an $\mathcal{ALC}$-concept description into NNF yields an equivalent concept description, TBox or ABox of the same size.

Let $\mathcal{A}_0$ be an $\mathcal{ALC}$-ABox that is to be tested for consistency. In a first preprocessing step the definitions from the TBox are expanded.[2] More precisely, names of defined concepts are replaced by the right-hand sides of their definitions in the TBox. This replacement is done exhaustively until only names of primitive concepts appear in the ABox $\mathcal{A}_0$. Next, this ABox is transformed into NNF. In order to test consistency of the normalized $\mathcal{A}_0$, the algorithm applies *tableau rules* to this ABox until no more rules apply. The tableau rules for $\mathcal{ALC}$ are depicted in Fig. 1. Tableau rules in general are consistency preserving transformation rules.

The tableau rule $\rightarrow_{\sqcup}$ that handles disjunction is *nondeterministic*. It transforms a given ABox into two new ABoxes such that the original ABox is consistent if *one* of the new ABoxes is so. For this reason, we will consider finite sets of ABoxes $\mathcal{S} = \{\mathcal{A}_1, \ldots, \mathcal{A}_k\}$ instead of single ones. Such a set of ABoxes is *consistent* iff there is

---

[2] Recall, that we are dealing with unfoldable TBoxes (Def. 6).

some i, $1 \leq i \leq k$, such that $\mathcal{A}_i$ is consistent. A tableau rule of Fig. 1 is applied to a given finite set of ABoxes $\mathcal{S}$ as follows: it takes an element $\mathcal{A}$ of $\mathcal{S}$, and replaces it by one ABox $\mathcal{A}'$ or, in case of $\rightarrow_{\sqcup}$ by two ABoxes $\mathcal{A}'$ and $\mathcal{A}''$.

Definition 13 (Clash, complete ABox, closed ABox). *An ABox $\mathcal{A}$ contains a* clash *iff* $\{A(x), \neg A(x)\} \subseteq \mathcal{A}$ *for some individual name x and some concept name A. An ABox $\mathcal{A}$ is called*

- complete *iff none of the tableau rules of Fig. 1 applies to it, and*
- closed *if it contains a clash, and* open *otherwise.*

The *consistency algorithm for $\mathcal{ALC}$* proceeds in the following steps. It starts with the singleton set of ABoxes $\{\mathcal{A}_0\}$, and applies the rules from Fig. 1 in arbitrary order until no more rules apply. The algorithm returns "consistent" if the set $\widehat{\mathcal{S}}$ of ABoxes obtained by exhaustively applying the tableau rules contains an open ABox, and "inconsistent" otherwise.

For this procedure, one can show that it is sound, complete and terminating by examining the individual tableau rules. For termination, it is easy to see that each rule application is monotonic in the sense that every rule application extends the number of concept assertions for the individuals in $\mathcal{A}$ and it never removes elements from $\mathcal{A}$. Furthermore, each concept description that appears in $\mathcal{A}$ due to the application of the tableau rules is a sub-concept description of a concept description that appears already in the initial ABox $\mathcal{A}_0$. These two facts together imply that the application of tableau rules terminates. Completeness of the procedure can easily be seen from the definition of a clash. Soundness can be shown by showing local correctness of the individual tableau rules. Local correctness means that the rules preserve consistency, i.e., if $\widehat{\mathcal{S}'}$ is obtained from the finite set of ABoxes $\widehat{\mathcal{S}}$ by application of a transformation rule, then $\widehat{\mathcal{S}}$ is consistent iff $\widehat{\mathcal{S}'}$ is consistent.

Due to space limitations, we refer the reader to [2, 6] for the proofs for soundness, completeness and termination of the tableau algorithm for $\mathcal{ALC}$.

For general TBoxes, the tableau algorithm needs to be extended by a rule for treating GCIs and a more complex mechanism to ensure termination. For a given general TBox $\mathcal{T} = \{C_1 \sqsubseteq D_1, \ldots, C_n \sqsubseteq D_n\}$, it is easy to see that the general TBox consisting of the single GCI of the form

$$\top \sqsubseteq (\neg C_1 \sqcup D_1) \sqcap \ldots \sqcap (\neg C_n \sqcup D_n)$$

is equivalent to $\mathcal{T}$, i.e., they have the same models. Thus, reasoning for general TBoxes can be done by taking a general TBox that consists of a single GCI of the form $\top \sqsubseteq C$, where C is a concept description constructed from the GCIs as above. This GCI states that every element in the model belongs to C. To capture this in the tableau method, we add a new rule: the $\rightarrow_{\top \sqsubseteq C}$-rule adds the concept assertion $C(x)$ in case the individual name x occurs in the ABox $\mathcal{A}$, and $C(x)$ is not yet present in $\mathcal{A}$. Local correctness, soundness, and completeness of this procedure can easily be shown. However, the procedure does not terminate, due to cyclic axioms. To regain termination, cyclic computations need to be detected and the application of the $\rightarrow_{\exists}$-rule must be blocked. For two individuals a and b, we say that a is *younger* than b, if a was introduced by an application of the $\rightarrow_\exists$-rule after b was already present in the ABox. The application of the $\rightarrow_\exists$-rule to an individual x is *blocked* by an individual y in an ABox $\mathcal{A}$ iff

- x is younger than y, and
- $\{C \mid C(x) \in \mathcal{A}\} \subseteq \{C \mid C(y) \in \mathcal{A}\}$.

The main idea underlying blocking is that the blocked individual x can use the role successors of y instead of generating new role successors.

The complexity of the consistency problem in $\mathcal{ALC}$ w.r.t. unfoldable TBoxes is PSpace-complete [92, 66]. In case general TBoxes are used, the complexity of testing consistency is ExpTime-complete [89]. For the DLs underlying the OWL standard the complexity of testing consistency is even higher. Reasoning in the DL underlying the OWL 1.0 standard $\mathcal{SHOIQ}$ is NExpTime-complete [98] and for the DL $\mathcal{SROIQ}$, which is the basis for the OWL 2 standard, it is even N2ExpTime [64].

3.2 Reasoning in $\mathcal{EL}$

Since the DL $\mathcal{EL}$ does neither offer negation nor the bottom concept, contradictions cannot be expressed and thus testing satisfiability is trivial in $\mathcal{EL}$. For testing subsumption in $\mathcal{EL}$, it was shown in [25] that reasoning can be done in polynomial time. This result was rather surprising. For the very similar DL $\mathcal{FL}_0$, which allows for value restrictions instead of existential restrictions, reasoning w.r.t. general TBoxes is ExpTime-complete [46]. For a collection of extensions of $\mathcal{EL}$ it was investigated, whether they have the same nice computational properties [26, 4, 5]. These investigations identified extensions of $\mathcal{EL}$ that allow for efficient classification. The DL $\mathcal{EL}^{++}$ extends $\mathcal{EL}$ with the bottom concept ($\bot$), nominals, a restricted form of concrete domains, and a restricted form of so-called role-value maps. For this DL, it was shown in [5] that almost all additions of other typical DL constructors to $\mathcal{EL}$ make subsumption w.r.t. general TBoxes ExpTime-complete. The DL $\mathcal{EL}^{++}$ is the closest DL to the OWL 2 EL profile.

Despite its limited expressivity, $\mathcal{EL}$ is highly relevant for practical applications. In fact, both the large medical ontology SNOMED CT[3] and the Gene Ontology[4] can be expressed in $\mathcal{EL}$.

3.3 Subsumption in $\mathcal{EL}$

The polynomial time algorithm for computing subsumption w.r.t. a general TBox actually performs classification of the whole TBox, i.e., it computes the subsumption relationships between all named concepts of a given TBox simultaneously. This algorithm proceeds in four steps:

1. Normalize the TBox.
2. Translate the normalized TBox into completion sets.
3. Complete these sets using completion rules.
4. Read off the subsumption relationships from the normalized graph.

---

[3] http://www.ihtsdo.org/snomed-ct/
[4] http://www.geneontology.org/

$$
\begin{array}{ll}
\textbf{NF1} & C \sqcap \hat{D} \sqsubseteq E \longrightarrow \{\, \hat{D} \sqsubseteq A, C \sqcap A \sqsubseteq E \,\} \\[4pt]
\textbf{NF2} & \exists r.\hat{C} \sqsubseteq D \longrightarrow \{\, \hat{C} \sqsubseteq A, \exists r.A \sqsubseteq D \,\} \\[4pt]
\textbf{NF3} & \hat{C} \sqsubseteq \hat{D} \longrightarrow \{\, \hat{C} \sqsubseteq A, A \sqsubseteq \hat{D} \,\} \\[4pt]
\textbf{NF4} & B \sqsubseteq \exists r.\hat{C} \longrightarrow \{\, B \sqsubseteq \exists r.A, A \sqsubseteq \hat{C} \,\} \\[4pt]
\textbf{NF5} & B \sqsubseteq C \sqcap D \longrightarrow \{\, B \sqsubseteq C, B \sqsubseteq D \,\}
\end{array}
$$

where $\hat{C}, \hat{D}$ are complex concept descriptions and A is a new concept name.

Fig. 2. $\mathcal{EL}$ normalization rules

The normal form for $\mathcal{EL}$-TBoxes required in the first step is defined as follows.

Definition 14 (Normal form for $\mathcal{EL}$-TBoxes). *An $\mathcal{EL}$-TBox $\mathcal{T}$ is in* normal form *if all concept inclusions have one of the following forms:*

$$
A_1 \sqsubseteq B, \quad A_1 \sqcap A_2 \sqsubseteq B, \quad A_1 \sqsubseteq \exists r.A_2 \quad \text{or} \quad \exists r.A_1 \sqsubseteq B,
$$

where A_1, A_2 and B are concept names appearing in $\mathcal{T}$ or the top-concept $\top$.

Any $\mathcal{EL}$-TBox $\mathcal{T}$ can be transformed into a normalized TBox $\mathcal{T}'$ by simply introducing new concept names. $\mathcal{EL}$-TBoxes can be transformed into normal form by applying the normalization rules displayed in Fig. 2 exhaustively. These rules replace the GCI on the left-hand side of the rule with the set of GCIs on the right-hand side of the rule. The idea behind the normalization rules is to introduce names for complex sub-concept descriptions. It suffices to obtain a TBox that is a subsumption-equivalent TBox to the original one, i.e., the original and the normalized TBox capture the same subsumption relationships for the named concepts from the original TBox. Thus it suffices to introduce the new concept names with GCIs instead of equivalences. The transformation into normal form can be done in linear time.

The completion algorithm works on a data-structure called *completion sets*. There are two kinds of completion sets used in the algorithm:

- $S(A)$ for each concept name A mentioned in the normalized TBox, and
- $S(A, r)$ for each concept name A and role name r mentioned in the normalized TBox.

Both kinds of completion sets contain concept names and $\top$. By $S_{\mathcal{T}}$ we denote the set containing all completion sets of the TBox $\mathcal{T}$. In the completion algorithm, the completion sets are initialized as follows:

- $S(A) := \{A, \top\}$ for each concept name A mentioned in the normalized TBox, and
- $S(A, r) := \emptyset$ for each concept name A and role name r mentioned in the normalized TBox.

CR1	If $C' \sqsubseteq D \in \mathcal{T}, C' \in S(C)$, and $D \notin S(C)$ then add D to $S(C)$.	
CR2	If $C_1 \sqcap C_2 \sqsubseteq D \in \mathcal{T}, C_1, C_2 \in S(C)$, and $D \notin S(C)$ then add D to $S(C)$.	
CR3	If $C' \sqsubseteq \exists r.D \in \mathcal{T}, C' \in S(C)$, and $D \notin S(C, r)$ then add D to $S(C, r)$.	
CR4	If $\exists r.D' \sqsubseteq E \in \mathcal{T}, D \in S(C, r), D' \in S(D)$, and $E \notin S(C)$ then add E to $S(C)$.	

Fig. 3. $\mathcal{EL}$ completion rules

The intuition is that the completion rules make implicit subsumption relationships explicit in the following sense:

- $B \in S(A)$ implies that $A \sqsubseteq_{\mathcal{T}} B$, i.e., $S(A)$ contains only subsumers of A, and
- $B \in S(A, r)$ implies that $A \sqsubseteq_{\mathcal{T}} \exists r.B$, i.e., $S(A, r)$ contains only concept names B s.t. A is subsumed by $\exists r.B$.

In fact, it can be shown that these properties of the completion sets are *invariants* and thus do not change during completion. Clearly, this holds for the initial elements of the completion. After initialization all completion sets in $\mathsf{S}_{\mathcal{T}}$ are extended by applying the completion rules that are shown in Fig. 3 exhaustively, i.e., until no more rule applies. It is easy to see that the rules preserve the above invariants. In each of the rules the last condition ensures that the rule is only applied once to the same concepts and completion sets. The first rule **CR1** propagates the transitivity of subsumption. The second **CR2** ensures that if a conjunction implies a concept C w.r.t. $\mathcal{T}$ and the conjuncts are already in the completion set of a concept, then C has to be in that completion set as well. The rule **CR3** is applicable if a concept name implies an existential restriction w.r.t. $\mathcal{T}$ and this concept name is contained in the completion set $S(C)$, then the existential restriction is implied by C as well. The most complicated rule is **CR4**. The axiom $\exists r.D' \sqsubseteq E \in \mathcal{T}$ implies $\exists r.D' \sqsubseteq_{\mathcal{T}} E$, and the assumption that the invariants are satisfied before applying the rule yields $D \sqsubseteq_{\mathcal{T}} D'$ and $C \sqsubseteq_{\mathcal{T}} \exists r.D$. The subsumption relationship $D \sqsubseteq_{\mathcal{T}} D'$ then implies $\exists r.D \sqsubseteq_{\mathcal{T}} \exists r.D'$. By applying transitivity of the subsumption relation $\sqsubseteq_{\mathcal{T}}$, we obtain $C \sqsubseteq_{\mathcal{T}} E$.

Once the completion process has terminated, the subsumption relation between two named concepts A and B can be tested by checking whether $B \in S(A)$. The fact that subsumption in $\mathcal{EL}$ w.r.t. general TBoxes can be decided in polynomial time follows from the following statements:

1. Rule application terminates after a polynomial number of steps.
2. If no more rules are applicable, then $A \sqsubseteq_{\mathcal{T}} B$ iff $B \in S(A)$.

The first statement holds, since the number of completion sets, of the kind $S(A)$ is linear in size of the TBox. In addition, the number of completion set of the kind $S(A, r)$ is quadratic in the size of $\mathcal{T}$. The size of the completion sets is bounded by the number of concept names and role names, and each rule application extends at least one label.

Theorem 1. *Subsumption in $\mathcal{EL}$ is polynomial w.r.t. general TBoxes.*

This nice computational property transfers also to $\mathcal{EL}^{++}$ [5], the DL corresponding closest to the OWL 2 EL profile.

The first implementation of the subsumption algorithm for $\mathcal{EL}$ sketched above is the CEL system [11,71]. This system showed that the classification of the very large knowledge bases can be done in runtime acceptable for practical applications. For instance, classifying the knowledge base SNOMED CT, which contains more than 300.000 axioms takes less than half an hour and classification of the Gene Ontology, which contains more than 20.000 axioms, takes only 6 seconds [12].

4   Explanation of Reasoning Results

DL knowledge bases often contain thousands of axioms and have a complex structure due to the use of GCIs. These knowledge bases are developed by users who are experts in the domain to be modeled, but have little expertise in knowledge representation or logic. For this sort of applications, it is necessary that the development process of the knowledge base is supported by automated services implemented in the DL system.

Classical DL reasoning systems can detect that a certain consequence holds, such as an inconsistency or a subsumption relation, but they give no evidence *why* it holds. The reasoning service explanation facilitates better understanding of the knowledge base and gives a starting point to resolve an unwanted consequence in the knowledge base. For instance, the SNOMED ontology contains the subsumption relation:

$$\text{Amputation-of-Finger} \sqsubseteq \text{Amputation-of-Arm}.$$

A user who wants to correct this, faces the task of finding the axioms responsible for this unintended subsumption relation among 350.000 others. Clearly, automated support is needed for this task. A first step towards providing such support was described in [90], where an algorithm for computing all minimal subsets of a given knowledge base that have a given consequence is described. This approach was extended to expressive DLs in [83].

For a TBox $\mathcal{T}$ and a consequence c an *explanation* points to the "source" of the consequence, which is a subset of $\mathcal{T}$ that contributes to the consequence c. We call a *minimal axiom set* (MinA) a minimal subset (w.r.t. size) of a TBox $\mathcal{T}$, that has a certain consequence. *Axiom pinpointing* is the process of computing MinAs.

Example 1. Consider the following TBox:

$$\mathcal{T}_{ex} = \{ \qquad \begin{array}{lr} \text{Cat} \sqsubseteq \exists \text{ has-parent. Cat}, & \text{I} \\ \text{Cat} \sqsubseteq \text{Pet}, & \text{II} \\ \exists \text{ has-parent.Pet} \sqsubseteq \text{Animal}, & \text{III} \\ \text{Pet} \sqsubseteq \text{Animal} \qquad \} & \text{IV} \end{array}$$

For the TBox $\mathcal{T}_{ex}$, we find the consequence Cat $\sqsubseteq_{\mathcal{T}_{ex}}$ Animal. The consequence holds since axiom I says that cats are pets and pets are in turn animals by axiom IV. This consequence also follows from $\mathcal{T}_{ex}$ by using axiom I and axiom II, which together say that a cat has a parent that is a pet. Now from this together with axiom III it, follows that cats are animals. Thus, the one consequence has several MinAs, namely: {I, IV} and {I, II, III}.

It turns out that there may be exponentially many MinAs, which shows that an algorithm for computing *all* MinAs needs exponential time in the size of the input TBox. In order to obtain an explanation for a consequence, we need to compute one single MinA of the consequence. There are two general approaches for pinpointing, i.e., computing a MinA of a consequence:

Black box approach, which uses a DL reasoner as an oracle, i.e, it repetitively queries the reasoner to compute a MinA.

Glass box approach, which modifies the internals of a DL reasoner s.t. it yields a MinA directly when computing an inference.

While the black box approach is independent of the reasoner, the glass box approach needs to be tailored to the reasoning method in use. We examine the black box approach first, which is the method of choice for expressive DLs, then we discuss the glass box approach for completion-based reasoning in $\mathcal{EL}$.

The task of computing explanations has also been considered in other research areas. For example, in the SAT community, people have considered the problem of computing minimally unsatisfiable subsets of a set of propositional formulae. Approaches for computing these sets developed there include algorithms that call a SAT solver as a black box [65, 20] but also algorithms that extend a resolution-based SAT solver directly [34, 103].

4.1   Black Box Method for Pinpointing

Assume we want to perform pinpointing for the consequence $A \sqsubseteq B$ w.r.t. the TBox $\mathcal{T}$. The basic idea underlying the black box method is a kind of uninformed search: Given a TBox $\mathcal{T}$ and the consequence $A \sqsubseteq B$: simply remove the first axiom from the TBox $\mathcal{T}$ and test whether the consequence still holds. If so, continue with the second axiom. If the consequence does *not* follow from the TBox with the first axiom removed, put the axiom back to the TBox and then test the second axiom. This naive method always performs as many subsumption tests as the number of axioms in the TBox. Since MinAs are often quite small, this is not a feasible method for very large TBoxes.

A more efficient method would not proceed axiom-wise, but first compute a not necessarily minimal subset of the TBox from which the consequence follows and then minimize this set using the naive procedure. This approach is only feasible if the algorithm for the first step produces fairly small sets of axioms and is efficient.

The black box method is independent of the DL in use and can be used to compute explanations for any DL, provided there is a DL reasoner for the DL and the consequence in question. This method can easily be implemented on top of a DL reasoner and does not require to change the internal structure of the reasoner. This is the reason why most implementations of pinpointing are based on the black box approach.

For $\mathcal{EL}$ the black box pinpointing algorithm has been implemented in the DL reasoning system CEL [16, 19, 97]. For a variant of the medical knowledge base GALEN [87] with 4000 axioms the overall run-time for computing a MinA with the non-naive method took 9:45 min. In contrast the naive method took seven hours for the same task. The first implementation of the black box method for pinpointing was done for the ontology editor SWOOP [62] based on the methods described in [83]. A more recent implementation of black box pinpointing was done in the ontology editor PROTÉGÉ. This implementation allows pinpointing even for *parts* of axioms that contribute to deriving a consequence [47].

4.2 Glass Box Pinpointing for $\mathcal{EL}$

The glass box approach for computing an explanation depends on the DL used and the reasoning method employed. It requires that the internals of a reasoner are modified by adding label sets to the reasoning procedure that collect the relevant axioms already during the computation of the consequence. For $\mathcal{EL}$, we modify the completion algorithm for subsumption from Section 3.3 to compute one explanation for a subsumption relationship. To this end, we annotate every element in the completion sets in S with a monotone Boolean formula that captures the MinAs.[5] The glass box algorithm for $\mathcal{EL}$ was described in [15] and extended in [16].

The *basic labeling* assigns to every GCI $t \in \mathcal{T}$ a unique propositional variable $lab(t)$ as a label. By $lab(\mathcal{T})$ we denote the set of all propositional variables labeling GCIs in the TBox $\mathcal{T}$. Now, a *monotone Boolean formula over $lab(\mathcal{T})$* is a Boolean formula using

- (some of) the variables in $lab(\mathcal{T})$, and
- only the connectives $\wedge$, $\vee$ and *true* for truth.

Its propositional *valuation* (denoted ν) is the set of propositional variables that make the formula true when they are assigned the value true. For a valuation $\nu \subseteq lab(\mathcal{T})$, let $\mathcal{T}_\nu := \{t \in \mathcal{T} \mid lab(\mathcal{T}) \in \nu\}$. The idea is that the valuation characterizes a combination of axiom labels. These labels are mapped back to the actual axioms from the TBox $\mathcal{T}$ by $\mathcal{T}_\nu$.

Definition 15 (Pinpointing formula). *Let $\mathcal{T}$ be an $\mathcal{EL}$-TBox and A and B concept names occurring in $\mathcal{T}$. The monotone Boolean formula ϕ over $lab(\mathcal{T})$ is a pinpointing formula for $\mathcal{T}$ w.r.t. $A \sqsubseteq_\mathcal{T} B$, if the following holds for every valuation $\nu \subseteq lab(\mathcal{T})$:*

$$A \sqsubseteq_{\mathcal{T}_\nu} B \text{ iff } \nu \text{ satisfies } \phi.$$

Consider Example 1 again. Take $lab(\mathcal{T}_{ex}) := \{\mathrm{I}, \mathrm{II}, \mathrm{III}, \mathrm{IV}\}$ as the set of propositional variables, then $\mathrm{II} \wedge (\mathrm{IV} \vee (\mathrm{I} \wedge \mathrm{III}))$ is a pinpointing formula for $\mathcal{T}_{ex}$ w.r.t. $A \sqsubseteq_{\mathcal{T}_{ex}} B$.

Lemma 1. *Let ϕ be a pinpointing formula for the TBox $\mathcal{T}$ w.r.t. $A \sqsubseteq_\mathcal{T} B$. If valuations are ordered by set inclusions, then*

$$M = \{\mathcal{T}_\nu \mid \nu \text{ is a minimal valuation satisfying } \phi\}$$

---

[5] This method for generating explanations was first applied for default reasoning in [8].

is the set of all *MinAs for* $\mathcal{T}$ *w.r.t.* $A \sqsubseteq_{\mathcal{T}} B$.

Proof. We need to show the following claims:

1. M contains only MinAs.
2. There is no MinA m_1 s.t. $m_1 \notin M$.

Show claim 1.:
For each set of axioms $m \in M$ there is a valuation ν_m s.t. $\nu_m = lab(m)$, which is minimal in size and that satisfies ϕ. Since ϕ is satisfied, $A \sqsubseteq_{\mathcal{T}} B$ holds. Since ν_m is minimal there is no subset of ν_m satisfying ϕ, and thus m is a MinA.

Show claim 2.:
Assume m_1 is a MinA for $\mathcal{T}$ w.r.t. $A \sqsubseteq_{\mathcal{T}} B$ and $m_1 \notin M$. Since m_1 is a MinA, m_1 is minimal and $A \sqsubseteq_{m_1} B$ holds. Let ν_{m_1} be the valuation $\nu_{m_1} = lab(m_1)$. From $A \sqsubseteq_{\mathcal{T}} B$ follows ν_{m_1} satisfies the pinpointing formula ϕ. Thus, m_1 induces a minimal valuation satisfying ϕ, which is a contradiction to $m_1 \notin M$. ❑

Lemma 1 guarantees that it is enough to compute the pinpointing formula to obtain *all* MinAs, i.e., explanations for the consequence in question. However, to obtain one MinA from the pinpointing formula, one can transform the pinpointing formula into disjunctive normal form, remove those disjuncts that are implied by other disjuncts and then pick one disjunct as the explanation.

Next, we describe the computation algorithm for pinpointing formulae in $\mathcal{EL}$ based on completion. Again, we want to explain $A \sqsubseteq B$ w.r.t. the $\mathcal{EL}$-TBox $\mathcal{T}$. Since the completion algorithm starts by normalizing the TBox, we need to introduce the labels for the original TBox and labels for the normalized TBox $\mathcal{T}'$ as well. The labels of the normalized TBox $\mathcal{T}'$ need to "keep track" of the corresponding axioms in the original TBox.

The completion procedure needs to be adapted to propagate the labels and to construct the pinpointing formula. To this end, each element of the completion sets, say $X \in S(A)$, is labelled with a monotone Boolean formula: $lab(A, X)$. The initial elements of the completions sets $A \in S(A)$ and $\top \in S(A)$ are labelled with *true*, i.e., $lab(A, A) = lab(A, \top) = true$ for all concept names appearing in $\mathcal{T}$. Now, we need to modify the completion rules from Fig. 3. Let the precondition of a completion rule CRi be satisfied for a set of completion sets $S_{\mathcal{T}'}$ w.r.t. the TBox $\mathcal{T}'$. The modified rule collects the labels of those GCIs and completion sets that make the rule CRi applicable. Let ϕ be the conjunction of :

- labels of GCIs in $\mathcal{T}'$ that appear in the precondition of CRi, and
- labels of elements in completion sets in $S_{\mathcal{T}'}$ that appear in the precondition of CRi.

The conjunction collected in ϕ needs to be propagated to the consequence of the rule CRi. If the completion set element in the consequence of CRi is *not* in $S_{\mathcal{T}'}$, then it is added with label ϕ. In case the consequence of CRi *is* already in $S_{\mathcal{T}'}$ and has the label ψ, the completion algorithm has derived the consequence again. In this case, ψ and ϕ are compared. If $\psi \wedge \phi \not\equiv \psi$, the consequence of CRi is derived in an alternative way and the label of this consequence is changed to $\phi \vee \psi$. The new label of the consequence is a more general Boolean formula. If $\psi \wedge \phi \equiv \psi$, then ϕ implies ψ. In this case the rule CRi is not applied.

Example 2. Consider Example 1 again. To compute the pinpointing formula for Cat $\sqsubseteq_{\mathcal{T}_{ex}}$ Animal, the set of completion sets $S_{\mathcal{T}_{ex}}$ is initialized as follows:

$$S_{\mathcal{T}_{ex}} = \{ \ (\text{Cat}, \top)^{true}, (\text{Cat}, \text{Cat})^{true},$$
$$(\text{Pet}, \top)^{true}, (\text{Pet}, \text{Pet})^{true},$$
$$(\text{Animal}, \top)^{true}, (\text{Animal}, \text{Animal})^{true} \ \}.$$

Then we can apply the modified rules:

- Using axiom II: Cat $\sqsubseteq$ Pet $\in \mathcal{T}_{ex}$ and $(\text{Cat}, \text{Cat})^{true} \in S_{\mathcal{T}_{ex}}$, add $(\text{Cat}, \text{Pet})^{\text{II} \wedge true}$ to $S_{\mathcal{T}_{ex}}$.
- Using axiom I: Cat $\sqsubseteq \exists$ has-parent. Cat $\in \mathcal{T}_{ex}$ and $(\text{Cat}, \text{Cat})^{true} \in S_{\mathcal{T}_{ex}}$, add $(\text{Cat}, \text{has-parent}, \text{Pet})^{\text{I} \wedge true}$ to $S_{\mathcal{T}_{ex}}$.
- Using axiom IV: Pet $\sqsubseteq$ Animal $\in \mathcal{T}_{ex}$ and $(\text{Cat}, \text{Pet})^{\text{II} \wedge true} \in S_{\mathcal{T}_{ex}}$, add $(\text{Cat}, \text{Animal})^{\text{II} \wedge \text{IV} \wedge true}$ to $S_{\mathcal{T}_{ex}}$.
- Using axiom III: $\exists$ has-parent.Pet $\sqsubseteq$ Animal $\in \mathcal{T}_{ex}$ and $\{(\text{Cat}, \text{Pet})^{\text{II} \wedge true}, (\text{Cat}, \text{has-parent}, \text{Pet})^{\text{I} \wedge true})\} \subset S_{\mathcal{T}_{ex}}$, modify $(\text{Cat}, \text{Animal})^{\text{II} \wedge \text{IV} \wedge true}$ to $(\text{Cat}, \text{Animal})^{(\text{II} \wedge \text{IV} \wedge true) \vee (\text{III} \wedge \text{II} \wedge \text{I} \wedge true)}$.

Now, $lab(\text{Cat}, \text{Animal}) = (\text{II} \wedge \text{IV}) \vee (\text{III} \wedge \text{II} \wedge \text{I})$ is the pinpointing formula for $\mathcal{T}_{ex}$ w.r.t. Cat $\sqsubseteq_{\mathcal{T}_{ex}}$ Animal.

The modified completion algorithm always terminates, but not necessarily in polynomial time due to the possibility of repeated generalization of the label. Testing equivalence of monotone Boolean formulae is an NP-complete problem. However, given formulae over n propositional variables whose size is exponential in n, equivalence can be tested in time exponential in n. Thus, there are at most exponentially many rule applications and each of them takes at most exponential time. This yields an exponential time bound for the execution of the pinpointing algorithm.

However, the set of completion sets S obtained by the described process is identical to the one obtained by the unmodified algorithm. After the modified completion algorithm has terminated, the label $lab(A, B)$ is a pinpointing formula for $\mathcal{T}$ w.r.t. $A \sqsubseteq_{\mathcal{T}} B$.

Theorem 2. *Given an $\mathcal{EL}$-TBox $\mathcal{T}$ in normal form, the pinpointing algorithm terminates in time exponential in the size of $\mathcal{T}$. After termination, the resulting set of completion sets $S_{\mathcal{T}}$ satisfies the following two properties for all concept names A, B occurring in $\mathcal{T}$:*

1. *$A \sqsubseteq_{\mathcal{T}} B$ iff $(S(A), B) \in S_{\mathcal{T}}$, and*
2. *$lab(A, B)$ is a pinpointing formula for $\mathcal{T}$ w.r.t $A \sqsubseteq_{\mathcal{T}} B$.*

This result was shown in [16] for the DL $\mathcal{EL}^{++}$. In the example, the TBox $\mathcal{T}_{ex}$ is already in normal form. In the general case, the TBox needs to be normalized and the pinpointing formula obtained by the modified completion needs to reconstruct the labels for the original axioms from the label of the normalized axioms.

The propositional variables from the normalized TBox in ϕ are replaced with those of the original one. More precisely, each label of a normalized GCI is replaced by the disjunction of its source GCIs. Once the de-normalized pinpointing formula is obtained, it is transformed into disjunctive normal form. One disjunct of this formula yields a MinA and thus an explanation of the consequence. To sum up, the *pinpointing extension* of the $\mathcal{EL}$ subsumption algorithm proceeds in the following steps:

1. Label all axioms in $\mathcal{T}$.
2. Normalize $\mathcal{T}$ according the rules from Fig. 2.
3. Label each axiom in the normalized TBox $\mathcal{T}'$ and keep the source GCI of every normalized GCI.
4. Apply the completion rules from Fig. 3 *modified* as described.
5. De-normalize the pinpointing formula.
6. Build the disjunctive normal form.
7. Pick one disjunct as explanation.

Note that the transformation into disjunctive normal form may cause an exponential blow-up, which means that, in some cases, the pinpointing formula provides us with a compact representation of the set of all MinAs. Also note that this blow-up is not in the size of the pinpointing formula but rather in the number of variables. Thus, if the size of the pinpointing formula is already exponential in the size of the TBox $\mathcal{T}$, computing all MinAs from it is still "only" exponential in the size of $\mathcal{T}$.

The glass box approach for pinpointing has also been investigated for more expressive DLs such as $\mathcal{ALC}$ in [72]. A more general view on tableaux and pinpointing was taken in [14].

We presented methods to obtain an explanation for a consequence. In order to actually repair a DL knowledge base, it is necessary to alleviate *all* causes of an unwanted consequence. In order to support users to repair a knowledge base, all MinAs need to be computed. The glass box method for $\mathcal{EL}$ computes all MinAs and can be employed for knowledge base repair directly. For the black box approach, a method for obtaining all MinAs is described in [90, 60]. This method computes the first MinA by the algorithms described above and then employs a method based on *hitting sets* to obtain the remaining MinAs.

The mechanism of pinpointing is not only useful for explanation or repair of DL knowledge bases. Access restrictions to knowledge bases can be supported as well [9]. If a user only has access to a part of the ontology, it is not obvious whether certain consequences can be accessed by the user as well. By computing all MinAs for the consequence, it can be tested whether the consequence follows from the accessible part alone. In that case access to the consequence does not violate the access restrictions.

Acknowledgement. This article is based on the Description Logic tutorial by the author, which she taught at the 2009 Masters Ontology Spring School organized by the Meraka Institute in Tshwane (Pretoria), South Africa and it is based on the course material by Franz Baader at the 2009 Reasoning Web Summer School, see [2]. The author would like to thank the anonymous reviewers and Marcel Lippmann for valuable comments on earlier versions of this paper.

References

1. Baader, F.: Description logic terminology. In: [6], pp. 485–495. Cambridge University Press, Cambridge (2003)
2. Baader, F.: Description logics. In: Tessaris, S., Franconi, E., Eiter, T., Gutierrez, C., Handschuh, S., Rousset, M.-C., Schmidt, R.A. (eds.) Reasoning Web. Semantic Technologies for Information Systems. LNCS, vol. 5689, pp. 1–39. Springer, Heidelberg (2009)

3. Baader, F., Bauer, A., Baumgartner, P., Cregan, A., Gabaldon, A., Ji, K., Lee, K., Rajarat-
nam, D., Schwitter, R.: A novel architecture for situation awareness systems. In: Giese, M.,
Waaler, A. (eds.) TABLEAUX 2009. LNCS, vol. 5607, pp. 77–92. Springer, Heidelberg
(2009)
4. Baader, F., Brandt, S., Lutz, C.: Pushing the $\mathcal{EL}$ envelope. In: Proc. of the 19th Int. Joint
Conf. on Artificial Intelligence (IJCAI-05), Edinburgh, UK. Morgan-Kaufmann Publishers,
San Francisco (2005)
5. Baader, F., Brandt, S., Lutz, C.: Pushing the $\mathcal{EL}$ envelope further. In: Clark, K., Patel-
Schneider, P.F. (eds.) Proc. of the OWLED Workshop (2008)
6. Baader, F., Calvanese, D., McGuinness, D., Nardi, D., Patel-Schneider, P. (eds.): The Description Logic Handbook: Theory, Implementation, and Applications. Cambridge University Press, Cambridge (2003)
7. Baader, F., Franconi, E., Hollunder, B., Nebel, B., Profitlich, H.-J.: An empirical analysis of
optimization techniques for terminological representation systems or: Making KRIS get a
move on. Applied Artificial Intelligence. Special Issue on Knowledge Base Management 4,
109–132 (1994)
8. Baader, F., Hollunder, B.: Embedding defaults into terminological knowledge representation formalisms. In: Proc. of the 3rd Int. Conf. on the Principles of Knowledge Representation and Reasoning (KR-92), pp. 306–317. Morgan Kaufmann, Los Altos (1992)
9. Baader, F., Knechtel, M., Peñaloza, R.: A generic approach for large-scale ontological reasoning in the presence of access restrictions to the ontology's axioms. In: Bernstein, A.,
Karger, D.R., Heath, T., Feigenbaum, L., Maynard, D., Motta, E., Thirunarayan, K. (eds.)
ISWC 2009. LNCS, vol. 5823, pp. 49–64. Springer, Heidelberg (2009)
10. Baader, F., Küsters, R., Molitor, R.: Computing least common subsumer in description logics with existential restrictions. In: Dean, T. (ed.) Proc. of the 16th Int. Joint Conf. on Artificial Intelligence (IJCAI-99), pp. 96–101. Morgan Kaufmann, Los Altos (1999)
11. Baader, F., Lutz, C., Suntisrivaraporn, B.: CEL—a polynomial-time reasoner for life science ontologies. In: Furbach, U., Shankar, N. (eds.) IJCAR 2006. LNCS (LNAI), vol. 4130,
pp. 287–291. Springer, Heidelberg (2006), http://lat.inf.tu-dresden.de/
systems/cel/
12. Baader, F., Lutz, C., Suntisrivaraporn, B.: Is tractable reasoning in extensions of the description logic $\mathcal{EL}$ useful in practice? Journal of Logic, Language and Information, Special
Issue on Method for Modality, M4M (2007)
13. Baader, F., Lutz, C., Turhan, A.-Y.: Small is again beautiful in description logics. KI –
Künstliche Intelligenz 24(1), 25–33 (2010)
14. Baader, F., Peñaloza, R.: Axiom pinpointing in general tableaux. Journal of Logic and Computation; Special Issue: Tableaux and Analytic Proof Methods 20(1), 5–34 (2010)
15. Baader, F., Peñaloza, R., Suntisrivaraporn, B.: Pinpointing in the description logic $\mathcal{EL}$. In:
Calvanese, D., Franconi, E., Haarslev, V., Lembo, D., Motik, B., Tessaris, S., Turhan, A.-Y.
(eds.) Proc. of the 2007 Description Logic Workshop (DL 2007), CEUR-WS (2007)
16. Baader, F., Peñaloza, R., Suntisrivaraporn, B.: Pinpointing in the description logic $\mathcal{EL}^+$. In:
Hertzberg, J., Beetz, M., Englert, R. (eds.) KI 2007. LNCS (LNAI), vol. 4667, pp. 52–67.
Springer, Heidelberg (2007)
17. Baader, F., Sattler, U.: An overview of tableau algorithms for description logics. Studia
Logica 69, 5–40 (2001)
18. Baader, F., Sertkaya, B., Turhan, A.-Y.: Computing the least common subsumer w.r.t. a
background terminology. Journal of Applied Logics (2007)
19. Baader, F., Suntisrivaraporn, B.: Debugging SNOMED CT using axiom pinpointing in the
description logic $\mathcal{EL}^+$. In: Proc. of the International Conference on Representing and Shar-
ing Knowledge Using SNOMED (KR-MED'08), Phoenix, Arizona (2008)

20. Bailey, J., Stuckey, P.J.: Discovery of minimal unsatisfiable subsets of constraints using hitting set dualization. In: Hermenegildo, M.V., Cabeza, D. (eds.) PADL 2004. LNCS, vol. 3350, pp. 174–186. Springer, Heidelberg (2005)
21. Blackburn, P., de Rijke, M., Venema, Y.: Modal Logic. Cambridge Tracts in Theoretical Computer Science. Cambridge University Press, Cambridge (2001)
22. Borgida, A., Patel-Schneider, P.F.: A semantics and complete algorithm for subsumption in the CLASSIC description logic. Journal of Artificial Intelligence Research 1, 277–308 (1994)
23. Brachman, R.J., Borgida, A., McGuinness, D.L., Alperin Resnick, L.: The CLASSIC knowledge representation system or KL-ONE: the next generation. Preprints of the Workshop on Formal Aspects of Semantic Networks, Two Harbors, Cal. (1989)
24. Brachman, R.J., Levesque, H.J.: Readings in Knowledge Representation. Morgan Kaufmann, Los Altos (1985)
25. Brandt, S.: Polynomial time reasoning in a description logic with existential restrictions, GCI axioms, and—what else? In: de Mantáras, R.L., Saitta, L. (eds.) Proc. of the 16th European Conf. on Artificial Intelligence (ECAI-04), pp. 298–302. IOS Press, Amsterdam (2004)
26. Brandt, S.: Reasoning in $\mathcal{ELH}$ w.r.t. general concept inclusion axioms. LTCS-Report LTCS-04-03, Chair for Automata Theory, Institute for Theoretical Computer Science, Dresden University of Technology, Germany (2004), http://lat.inf.tu-dresden.de/research/reports.html
27. Brandt, S., Küsters, R., Turhan, A.-Y.: Approximation and difference in description logics. In: Fensel, D., McGuinness, D., Williams, M.-A. (eds.) Proc. of the 8th Int. Conf. on the Principles of Knowledge Representation and Reasoning (KR-02). Morgan Kaufmann Publishers, San Francisco (2002)
28. Calvanese, D., De Giacomo, G., Lembo, D., Lenzerini, M., Rosati, R.: DL-Lite: Tractable description logics for ontologies. In: Veloso, M.M., Kambhampati, S. (eds.) Proc. of the 20th Nat. Conf. on Artificial Intelligence (AAAI'05), pp. 602–607. AAAI Press/The MIT Press (2005)
29. Calvanese, D., De Giacomo, G., Lembo, D., Lenzerini, M., Rosati, R.: Data complexity of query answering in description logics. In: Proc. of the 10th Int. Conf. on the Principles of Knowledge Representation and Reasoning (KR 2006), pp. 260–270 (2006)
30. Calvanese, D., De Giacomo, G., Lembo, D., Lenzerini, M., Rosati, R.: Tractable reasoning and efficient query answering in description logics: The DL-Lite family. Journal of Automated Reasoning 39(3), 385–429 (2007)
31. Calvanese, D., Giacomo, G.D.: Expressive description logics. In: [6], pp. 178–218. Cambridge University Press, Cambridge (2003)
32. Cuenca Grau, B., Horrocks, I., Kazakov, Y., Sattler, U.: Modular reuse of ontologies: Theory and practice. Journal of Artificial Intelligence Research 31, 273–318 (2008)
33. Cuenca Grau, B., Kazakov, Y., Horrocks, I., Sattler, U.: A logical framework for modular integration of ontologies. In: Proc. of the 20th Int. Joint Conf. on Artificial Intelligence (IJCAI-07), pp. 298–303 (2007)
34. Davydov, G., Davydova, I., Büning, H.K.: An efficient algorithm for the minimal unsatisfiability problem for a subclass of cnf. Ann. Math. Artif. Intell. 23(3-4), 229–245 (1998)
35. Donini, F.M., Colucci, S., Di Noia, T., Di Sciascio, E.: A tableaux-based method for computing least common subsumers for expressive description logics. In: Proc. of the Twenty-First International Joint Conference on Artificial Intelligence (IJCAI-09), pp. 739–745. AAAI, Menlo Park (July 2009)
36. Eiter, T., Lutz, C., Ortiz, M., Simkus, M.: Query answering in description logics with transitive roles. In: Proc. of the 21st International Joint Conference on Artificial Intelligence IJCAI'09. AAAI Press, Menlo Park (2009)

37. Fitting, M.: Basic modal logic. In: Handbook of Logic in Artificial Intelligence and Logic Programming, vol. 1, pp. 365–448. Oxford Science Publications, Oxford (1993)
38. Ghilardi, S., Lutz, C., Wolter, F.: Did I damage my ontology? a case for conservative extensions in description logics. In: Doherty, P., Mylopoulos, J., Welty, C. (eds.) Proc. of the 10th Int. Conf. on the Principles of Knowledge Representation and Reasoning (KR-06), pp. 187–197. AAAI Press, Menlo Park (2006)
39. Glimm, B., Horrocks, I., Lutz, C., Sattler, U.: Conjunctive query answering for the description logic $\mathcal{SHIQ}$. In: Veloso, M.M. (ed.) Proc. of the 20th Int. Joint Conf. on Artificial Intelligence (IJCAI-07), pp. 399–404 (2007)
40. Grädel, E.: On the restraining power of guards. Journal of Symbolic Logic 64, 1719–1742 (1999)
41. Grädel, E., Kolaitis, P.G., Vardi, M.Y.: On the decision problem for two-variable first-order logic. Bulletin of Symbolic Logic 3(1), 53–69 (1997)
42. Haarslev, V., Möller, R.: High performance reasoning with very large knowledge bases: A practical case study. In: Nebel, B. (ed.) Proc. of the 17th Int. Joint Conf. on Artificial Intelligence (IJCAI-01), pp. 161–166 (2001)
43. Haarslev, V., Möller, R.: RACER system description. In: Goré, R., Leitsch, A., Nipkov, T. (eds.) IJCAR 2001. LNCS (LNAI), vol. 2083, p. 701. Springer, Heidelberg (2001)
44. Haarslev, V., Möller, R.: Optimization techniques for retrieving resources described in OWL/RDF documents: First results. In: Proc. of the 9th Int. Conf. on the Principles of Knowledge Representation and Reasoning (KR-04), pp. 163–173 (2004)
45. Haarslev, V., Möller, R., Turhan, A.-Y.: Exploiting pseudo models for TBox and ABox reasoning in expressive description logics. In: Goré, R., Leitsch, A., Nipkov, T. (eds.) IJCAR 2001. LNCS (LNAI), vol. 2083, p. 61. Springer, Heidelberg (2001)
46. Hofmann, M.: Proof-theoretic approach to description-logic. In: Panangaden, P. (ed.) Proc. of the 20th Ann. IEEE Symp. on Logic in Computer Science (LICS-05), pp. 229–237. IEEE Computer Society Press, Los Alamitos (2005)
47. Horridge, M., Parsia, B., Sattler, U.: Laconic and precise justifications in OWL. In: ISWC'08 The International Semantic Web Conference 2008, Karlsruhe, Germany (2008)
48. Horrocks, I.: Optimising Tableaux Decision Procedures for Description Logics. PhD thesis, University of Manchester (1997)
49. Horrocks, I.: Using an expressive description logic: FaCT or fiction? In: Cohn, A., Schubert, L., Shapiro, S. (eds.) Proc. of the 6th Int. Conf. on the Principles of Knowledge Representation and Reasoning (KR-98), pp. 636–647 (1998)
50. Horrocks, I.: Reasoning with expressive description logics: Theory and practice. In: Voronkov, A. (ed.) CADE 2002. LNCS (LNAI), vol. 2392, pp. 1–15. Springer, Heidelberg (2002)
51. Horrocks, I., Kutz, O., Sattler, U.: The even more irresistible $\mathcal{SROIQ}$. In: Doherty, P., Mylopoulos, J., Welty, C. (eds.) Proc. of the 10th Int. Conf. on the Principles of Knowledge Representation and Reasoning (KR-06), pp. 57–67. AAAI Press, Menlo Park (2006)
52. Horrocks, I., Patel-Schneider, P.: Optimizing description logic subsumption. Journal of Logic and Computation 9(3), 267–293 (1999)
53. Horrocks, I., Patel-Schneider, P., van Harmelen, F.: From $\mathcal{SHIQ}$ and RDF to OWL: The making of a web ontology language. Journal of Web Semantics 1(1), 7–26 (2003)
54. Horrocks, I., Sattler, U.: A description logic with transitive and inverse roles and role hierarchies. Journal of Logic and Computation 9(3), 385–410 (1999)
55. Horrocks, I., Sattler, U.: Optimised reasoning for $\mathcal{SHIQ}$. In: Proc. of the 15th European Conference on Artificial Intelligence (2002)
56. Horrocks, I., Sattler, U.: A tableaux decision procedure for $\mathcal{SHOIQ}$. In: Proc. of the 19th Int. Joint Conf. on Artificial Intelligence (IJCAI-05). Morgan Kaufmann, San Francisco (January 2005)

57. Horrocks, I., Sattler, U.: A tableau decision procedure for $\mathcal{SHOIQ}$. J. of Automated Reasoning 39(3), 249–276 (2007)
58. Horrocks, I., Sattler, U., Tobies, S.: Practical reasoning for very expressive description logics. J. of the Interest Group in Pure and Applied Logic 8(3), 239–264 (2000)
59. Hustadt, U., Schmidt, R.A., Georgieva, L.: A survey of decidable first-order fragments and description logics. Journal of Relational Methods in Computer Science 1, 251–276 (2004)
60. Kalyanpur, A., Parsia, B., Horridge, M., Sirin, E.: Finding all justifications of owl dl entailments. In: Aberer, K., Choi, K.-S., Noy, N., Allemang, D., Lee, K.-I., Nixon, L.J.B., Golbeck, J., Mika, P., Maynard, D., Mizoguchi, R., Schreiber, G., Cudré-Mauroux, P. (eds.) ASWC 2007 and ISWC 2007. LNCS, vol. 4825, pp. 267–280. Springer, Heidelberg (2007)
61. Kalyanpur, A., Parsia, B., Sirin, E., Cuenca Grau, B.: Repairing unsatisfiable concepts in owl ontologies. In: Sure, Y., Domingue, J. (eds.) ESWC 2006. LNCS, vol. 4011, pp. 170–184. Springer, Heidelberg (2006)
62. Kalyanpur, A., Parsia, B., Sirin, E., Cuenca Grau, B., Hendler, J.: Swoop: A web ontology editing browser. J. Web Sem. 4(2), 144–153 (2006)
63. Kalyanpur, A., Parsia, B., Sirin, E., Hendler, J.A.: Debugging unsatisfiable classes in OWL ontologies. J. Web Sem. 3(4), 268–293 (2005)
64. Kazakov, Y.: $\mathcal{RIQ}$ and $\mathcal{SROIQ}$ are harder than $\mathcal{SHOIQ}$. In: Brewka, G., Lang, J. (eds.) Proc. of the 11th Int. Conf. on the Principles of Knowledge Representation and Reasoning (KR-08), pp. 274–284. AAAI Press, Menlo Park (2008)
65. Liffiton, M.H., Sakallah, K.A.: Algorithms for computing minimal unsatisfiable subsets of constraints. J. Autom. Reasoning 40(1), 1–33 (2008)
66. Lutz, C.: In: Ganzinger, H., McAllester, D., Voronkov, A. (eds.) LPAR 1999. LNCS, vol. 1705, Springer, Heidelberg (1999)
67. Lutz, C.: The complexity of conjunctive query answering in expressive description logics. In: Armando, A., Baumgartner, P., Dowek, G. (eds.) IJCAR 2008. LNCS (LNAI), vol. 5195, pp. 179–193. Springer, Heidelberg (2008)
68. Lutz, C., Sattler, U., Wolter, F.: Description logics and the two-variable fragment. In: McGuiness, D., Pater-Schneider, P., Goble, C., Möller, R. (eds.) Proc. of the 2001 International Workshop in Description Logics (DL-2001), Stanford, California, USA, pp. 66–75 (2001)
69. Lutz, C., Walther, D., Wolter, F.: Conservative extensions in expressive description logics. In: Proc. of the 20th Int. Joint Conf. on Artificial Intelligence (IJCAI-07). AAAI Press, Menlo Park (2007)
70. Lutz, C., Wolter, F.: Deciding inseparability and conservative extensions in the description logic $\mathcal{EL}$. Journal of Symbolic Computation 45(2), 194–228 (2010)
71. Mendez, J., Suntisrivaraporn, B.: Reintroducing cel as an owl 2 el reasoner. In: Cuenca Grau, B., Horrocks, I., Motik, B., Sattler, U. (eds.) Proc. of the 2008 Description Logic Workshop (DL 2009), vol. 477. CEUR-WS (2009)
72. Meyer, T., Lee, K., Booth, R., Pan, J.Z.: Finding maximally satisfiable terminologies for the description logic $\mathcal{ALC}$. In: Proc. of the 21st Nat. Conf. on Artificial Intelligence (AAAI'06), AAAI Press/The MIT Press (2006)
73. Minsky, M.: A framework for representing knowledge. Technical report, MIT-AI Laboratory, Cambridge, MA, USA (1974)
74. Motik, B.: Reasoning in Description Logics using Resolution and Deductive Databases. PhD thesis, Universität Karlsruhe (2006)
75. Motik, B., Cuenca Grau, B., Horrocks, I., Wu, Z., Fokoue, A., Lutz, C.: Owl 2 web ontology language profiles. W3C Recommendation (October 27, 2009), http://www.w3.org/2009/REC-owl2-profiles-20091027/

76. Motik, B., Sattler, U.: A Comparison of Techniques for Querying Large Description Logic ABoxes. In: Hermann, M., Voronkov, A. (eds.) LPAR 2006. LNCS (LNAI), vol. 4246, pp. 227–241. Springer, Heidelberg (2006), http://kaon2.semanticweb.org/
77. Motik, B., Shearer, R., Horrocks, I.: A hypertableau calculus for $\mathcal{SHIQ}$. In: Calvanese, D., Franconi, E., Haarslev, V., Lembo, D., Motik, B., Tessaris, S., Turhan, A.-Y. (eds.) Proc. of the 2007 Description Logic Workshop, DL 2007 (2007)
78. Motik, B., Shearer, R., Horrocks, I.: Optimized Reasoning in Description Logics using Hypertableaux. In: Pfennig, F. (ed.) CADE-23. LNCS (LNAI), pp. 67–83. Springer, Heidelberg (2007)
79. Nebel, B.: Computational complexity of terminological reasoning in BACK. Artificial Intelligence Journal 34(3), 371–383 (1988)
80. Nebel, B.: Terminological reasoning is inherently intractable. Artificial Intelligence Journal 43, 235–249 (1990)
81. Nebel, B., von Luck, K.: Hybrid reasoning in BACK. In: Proc. of the 3rd Int. Sym. on Methodologies for Intelligent Systems (ISMIS-88), pp. 260–269. North-Holland Publ. Co., Amsterdam (1988)
82. Ortiz, M., Calvanese, D., Eiter, T.: Data complexity of query answering in expressive description logics via tableaux. Journal of Automated Reasoning 41(1), 61–98 (2008)
83. Parsia, B., Sirin, E., Kalyanpur, A.: Debugging OWL Ontologies. In: Ellis, A., Hagino, T. (eds.) Proc. of the 14th Int. World Wide Web Conference (WWW 2005), Japan, pp. 633–640 (2005)
84. Patel-Schneider, P., McGuinness, D.L., Brachman, R.J., Resnick, L.A., Borgida, A.: The CLASSIC knowledge representation system: Guiding principles and implementation rational. SIGART Bulletin 2(3), 108–113 (1991)
85. Quillian, M.R.: Word concepts: A theory and simulation of some basic capabilities. Behavioral Science 12, 410–430 (1967) Republished in [24]
86. Racer Systems GmbH & Co. KG. RacerPro Reference Manual Version 1.9 (December 2005), http://www.racer-systems.com/products/racerpro/reference-manual-1-9.pdf
87. Rector, A., Horrocks, I.: Experience building a large, re-usable medical ontology using a description logic with transitivity and concept inclusions. In: Proc. of the Workshop on Ontological Engineering. AAAI Spring Symposium, AAAI'97 (1997)
88. Sattler, U., Calvanese, D., Molitor, R.: Relationships with other formalisms. In: [6], pp. 137–177. Cambridge University Press, Cambridge (2003)
89. Schild, K.: A correspondence theory for terminological logics: preliminary report. In: Mylopoulos, J., Reiter, R. (eds.) Proc. of the 12th Int. Joint Conf. on Artificial Intelligence (IJCAI-91), Sydney, Australia (1991)
90. Schlobach, S., Cornet, R.: Non-standard reasoning services for the debugging of description logic terminologies. In: Gottlob, G., Walsh, T. (eds.) Proc. of the 18th Int. Joint Conf. on Artificial Intelligence (IJCAI-03), Mexico, pp. 355–362. Morgan Kaufmann, Los Altos (2003)
91. Schmidt-Schauß, M., Smolka, G.: Attributive concept descriptions with unions and complements. Technical Report SR-88-21, Deutsches Forschungszentrum für Künstliche Intelligenz (DFKI), Kaiserslautern, Germany (1988)
92. Schmidt-Schauß, M., Smolka, G.: Attributive concept descriptions with complements. Artificial Intelligence Journal 48(1), 1–26 (1991)
93. Sirin, E., Parsia, B.: Pellet system description. In: Parsia, B., Sattler, U., Toman, D. (eds.) Description Logics, CEUR Workshop Proceedings, vol. 189. CEUR-WS.org (2006)
94. Sowa, J.F. (ed.): Principles of Semantic Networks. Morgan Kaufmann, Los Altos (1991)
95. Sowa, J.F.: Semantic Networks. In: Encyclopedia of Artificial Intelligence. John Wiley & Sons, New York (1992)

96. Springer, T., Turhan, A.-Y.: Employing description logics in ambient intelligence for modeling and reasoning about complex situations. Journal of Ambient Intelligence and Smart Environments 1(3), 235–259 (2009)
97. Suntisrivaraporn, B.: Polynomial-Time Reasoning Support for Design and Maintenance of Large-Scale Biomedical Ontologies. PhD thesis, Fakultät Informatik, TU Dresden (2009), http://lat.inf.tu-dresden.de/research/phd/#Sun-PhD-2008
98. Tobies, S.: The complexity of reasoning with cardinality restrictions and nominals in expressive description logics. Journal of Artificial Intelligence Research 12, 199–217 (2000)
99. Tsarkov, D., Horrocks, I.: FaCT++ description logic reasoner: System description. In: Proc. of the 3rd Int. Joint Conf. on Automated Reasoning, IJCAR-06 (2006), FaCT++ download page, http://owl.man.ac.uk/factplusplus/
100. Tsarkov, D., Horrocks, I., Patel-Schneider, P.F.: Optimising terminological reasoning for expressive description logics. Journal of Automated Reasoning (2007)
101. Turhan, A.-Y.: On the Computation of Common Subsumers in Description Logics. PhD thesis, TU Dresden, Institute for Theoretical Computer Science (2007)
102. W3C OWL Working Group. Owl 2 web ontology language document overview. W3C Recommendation (October 27, 2009), http://www.w3.org/TR/2009/REC-owl2-overview-20091027/
103. Zhang, L., Malik, S.: Validating sat solvers using an independent resolution-based checker: Practical implementations and other applications. In: Design, Automation and Test in Europe Conference and Exposition (DATE 2003), Munich, Germany, March 3-7, pp. 10880–10885. IEEE Computer Society, Los Alamitos (2003)

Hybrid Reasoning with Non-monotonic Rules

Włodzimierz Drabent[1,2]

[1] Institute of Computer Science, Polish Academy of Sciences,
ul. Ordona 21, PL – 01-237 Warszawa, Poland
[2] Department of Computer and Information Science,
Linköping University, SE – 581 83 Linköping, Sweden
wdr@ida.liu.se

Abstract. This is an introduction to integrating logic programs with first order theories. The main motivation are the needs of Semantic Web to combine reasoning based on rule systems with that based on Description Logics (DL). We focus on approaches which are able to re-use existing reasoners (for DL and for rule systems). A central issue of this paper is non-monotonic reasoning, which is possibly the main feature of rule based reasoning absent in DL.

We discuss the main approaches to non-monotonic reasoning in logic programming. Then we show and classify various ways of integrating them with first order theories. We argue that for practical purposes none of the approaches seems sufficient, and an approach combining the features of so-called tight and loose coupling is needed.

1 Introduction

This paper is an introduction to combining logic programs with first order theories. The main motivation are the needs of Semantic Web to integrate reasoning based on rule systems with that based on Description Logics (DL). The former are dealt with by W3C RIF (Rule Interchange Format, under development), the latter are the basis of Web ontology languages OWL and OWL 2. A central issue of this presentation is non-monotonic reasoning, which is possibly the main feature of rule based reasoning absent in DL.

We first discuss non-monotonic reasoning in Logic Programming (LP), in other words the issue of negation in LP. We take care to compare the three main semantic approaches. Then we discuss and classify various ways of integrating first order theories and logic programs (with negation). In the intended applications the first order theories are description logics. We are interested in so-called heterogeneous, or hybrid, approaches. They make it possible to employ existing reasoners for LP and for first order theories. So we do not discuss the approaches for which integration is obtained by expressing both logic programs and DL in some special common formalism.

We try to avoid unnecessary formalities and to base our presentation on examples. On the other hand we want to keep the necessary mathematical rigour. The presentation is not comprehensive, for some technical details the reader is referred to other sources.

U. Aßmann, A. Bartho, and C. Wende (Eds.): Reasoning Web 2010, LNCS 6325, pp. 28–61, 2010.
© Springer-Verlag Berlin Heidelberg 2010

In Section 3 we present and compare the main semantic approaches to negation in Logic Programming. Section 4 presents and compares methods of combining logic programs with external first order theories.

2 Preliminaries

2.1 First Order Logic

Here we summarize main notions of first order logic (FOL). Its alphabet consists of disjoint alphabets of predicate symbols $\mathcal{P}$, function symbols $\mathcal{F}$ (including constants), variables $\mathcal{V}$, logical connectives $\{\neg, \wedge, \vee, \leftarrow, \leftrightarrow\}$, quantifiers $\{\exists, \forall\}$, and auxiliary symbols (parentheses etc). To each symbol from $\mathcal{P}$ or $\mathcal{F}$ there corresponds a natural number – its arity (the number of arguments). A constant is a symbol from $\mathcal{F}$ with arity 0. Sometimes we write f/n to state that the arity of f is n. A *term* is defined recursively as a variable, a constant, or an expression $f(t_1, \ldots, t_n)$ where $f \in \mathcal{F}$ has arity n and $t_1, \ldots, t_n$ are terms. An *atom* (or atomic formula) is $p(t_1, \ldots, t_n)$, where $t_1, \ldots, t_n$ are terms and $p \in \mathcal{P}$ has arity n. Formulas are built out of atoms in a standard way, using the connectives and quantifiers. A formula, or a term is *ground* if it does not contain variables.

Following the usual notation of logic programming, we often denote conjunction by a comma instead of $\wedge$, and use upper case letters for variables. Some parentheses in formulae may be skipped, assuming priorities of connectives in the order $\neg, \wedge, \vee, \leftarrow, \leftrightarrow$. So for instance $p \leftarrow q, \neg r \vee s$ stands for $p \leftarrow (q \wedge (\neg r)) \vee s)$.

We employ the standard model-theoretic semantics of FOL. An *interpretation* $\mathcal{I}$ consists of a nonempty set D called the *domain* of $\mathcal{I}$, and two functions assigning meaning to function and predicate symbols. To each function symbol $f \in \mathcal{F}$ of arity n an n-argument function $f^{\mathcal{I}} : D^n \to D$ is assigned. In particular, for a constant $c \in \mathcal{F}$ the corresponding (0 argument) function $c^{\mathcal{I}}$ is an element of D. To each predicate symbol $p \in \mathcal{P}$ of arity n there corresponds an n-argument relation $p^{\mathcal{I}} \subseteq D^n$. (In the special case of $n = 0$, D^0 is a one element set $\{\langle\rangle\}$.) We will use a logical constant (i.e. a predicate symbol of arity 0) *false* $\in \mathcal{P}$, and assume that it is false in any interpretation: $false^{\mathcal{I}} = \emptyset$.

A *variable assignment* is a function $\nu : \mathcal{V} \to D$. The *value* $t^{\mathcal{I},\nu}$ of a term t under $\mathcal{I}$ and ν is defined recursively: $v^{\mathcal{I},\nu} = \nu(v)$ for a variable v, and $(f(t_1, \ldots, t_n))^{\mathcal{I},\nu} = f^{\mathcal{I}}(t_1^{\mathcal{I},\nu}, \ldots, t_n^{\mathcal{I},\nu})$ for a term $f(t_1, \ldots, t_n)$, $n \geq 0$. An atom $p(t_1, \ldots, t_n)$ is true in $\mathcal{I}$ and ν (denoted by $\mathcal{I}, \nu \models p(t_1, \ldots, t_n)$) when $\langle t_1^{\mathcal{I},\nu}, \ldots, t_n^{\mathcal{I},\nu} \rangle \in p^{\mathcal{I}}$. Truth of non-atomic formulae is defined in a standard way; here we only present the definition for universally quantified formulae. Let X be a variable and F a formula. Then $\mathcal{I}, \nu \models \forall X F$ when $\mathcal{I}, \nu' \models F$ for each ν' such that $\nu'(Y) = \nu(Y)$ for every variable Y distinct from X.

A formula F is true in $\mathcal{I}$ (denoted by $\mathcal{I} \models F$) if $\mathcal{I}, \nu \models F$ for every ν. Thus $\mathcal{I} \models \forall X F$ iff (if and only if) $\mathcal{I} \models F$.

A first order (FO) theory is a set of formulas. A *model* of a theory T is an interpretation $\mathcal{I}$ such that $\mathcal{I} \models F$ for each formula $F \in T$. A formula F is a *logical consequence* of a theory T (written $T \models F$) if F is true in every model of T.[1]

Description logics (DL) are certain subsets of FOL. For the purposes of this paper we do not need to specify any actual DL. We will avoid the special syntax of DL. Instead we will use the standard syntax of FOL.

2.2   Logic Programming

In this section we briefly summarize main concepts of logic programming (LP). For a more complete introduction the reader is referred to textbooks and tutorials, like [56,3,4]. Our summary includes the semantics of positive programs. The semantics of programs with negation is dealt with in Section 3.

A *literal* is an atom, or a formula $\neg A$, where A is an atom; the latter is called a *negative literal*. A *clause* (or a *rule*) is a formula of the form

$$H \leftarrow B_1, \ldots, B_n \qquad\qquad (n \geq 0),$$

where H is an atom and $B_1, \ldots, B_n$ are literals. If H is $p(t_1, \ldots, t_m)$ then the clause is called a *clause for p*. If $B_1, \ldots, B_n$ are atoms then the clause is a *definite clause*. H is the *head* of the clause, $B_1, \ldots, B_n$ its *body*. When the body is empty ($n = 0$), the clause is *unary*; we often write unary clauses without the arrow. A set of definite clauses is called a *positive* (or *definite*) *program*. A set of clauses is called a *general program* (or *program with negation*, or simply *program*). When the set $\mathcal{F}$ of function symbols is finite and contains only constants, a program is called a (positive or general) *Datalog* program. (Sometimes additional conditions are imposed on variable occurrences in Datalog clauses, we do not use them here.) A *disjunctive clause* is of the form $H_1 \lor \ldots \lor H_m \leftarrow B_1, \ldots, B_n$, where $H_1, \ldots, H_n$ are atoms, $B_1, \ldots, B_n$ are literals, and $n, m \geq 0$.

When writing programs we will often use semicolons to separate clauses (as commas may occur within a clause).

A *substitution* is a mapping θ from variables to terms, such that the set $dom(\theta) = \{\, X \in \mathcal{V} \mid \theta(X) \neq X \,\}$ is finite. The result $E\theta$ of applying a substitution θ to an expression (i.e. a term or a formula) E is defined in a standard way. $E\theta$ is called an *instance* of E. A substitution η which maps distinct variables into distinct variables is called a *renaming substitution* (in other words, $\eta : \mathcal{V} \to \mathcal{V}$ and η is 1-1 and onto). For any such η, the instance $E\eta$ of E is called a *renaming* of E. The notions of a unifier and a most general unifier are defined in a standard way.

Declarative semantics. When dealing with logic programs we will often restrict the alphabet $\mathcal{P}$ of predicate symbols to those occurring in the program.

---

[1] Notice that $T \models \forall X F$ iff $T \models F$. So within a theory a formula can be replaced by its universal quantification: $T \cup \{F\} \models F'$ iff $T \cup \{\forall X F\} \models F'$ iff $T \cup \{\forall F\} \models F'$; where $\forall F$ stands for the universal closure $\forall X_1 \ldots \forall X_k F$, with $X_1, \ldots, X_n$ being the variables occurring in F.

A *Herbrand universe* is the set $\mathcal{HU}$ of ground terms (over a given alphabet). A *Herbrand base* is the set $\mathcal{HB}$ of ground atoms. An interpretation $\mathcal{I}$ whose domain is $\mathcal{HU}$, and in which $t^{\mathcal{I}} = t$ for any ground term t, is called a *Herbrand interpretation*. Herbrand interpretations are characterized by their treatment of predicate symbols (as the meaning of function symbols is fixed). A Herbrand interpretation $\mathcal{I}$ can be uniquely represented by a subset of $\mathcal{HB}$:

$$\{\, p(t_1, \ldots, t_n) \in \mathcal{HB} \mid \mathcal{I} \models p(t_1, \ldots, t_n)\,\}$$

We will usually not distinguish the interpretation from the set.

The semantics of positive programs is based on their logical consequences. An atom A is a logical consequence of a positive program P ($P \models A$) iff it is the root of a proof tree. A *proof tree* for P is a finite tree whose nodes are atoms, and whenever $B_1, \ldots, B_n$ ($n \geq 0$) are the children of a node B then $B \leftarrow B_1, \ldots, B_n$ is an instance of a clause of P. In particular, each leaf of a proof tree is an instance of a unary clause of the program.

The *least Herbrand model* of a positive program P is the Herbrand interpretation given by the set of ground atomic logical consequences of P:

$$\mathcal{M}_P = \{\, A \in \mathcal{HB} \mid P \models A\,\}.$$

$\mathcal{M}_P$ is indeed a model of P. It is a subset of any other Herbrand model of P, hence it is the least model under $\subseteq$ as the ordering. $\mathcal{M}_P$ is the set of the roots of ground proof trees for P. (This fact is usually expressed equivalently by employing the so-called immediate consequence operator T_P, and its least fixed point.) Least Herbrand models provide a useful characterization of the semantics of positive programs.

Operational semantics. The semantics discussed above is called the *declarative semantics*. Programs were treated as sets of logical axioms, and the issue was which formulae were their consequences. On the other hand, *operational semantics* deals with computing the consequences. The standard approach is called SLD-resolution. We outline it very briefly, referring the reader to textbooks for explanations.

SLD-resolution deals with *queries*, which are conjunctions of atoms. Given a query $Q_i = A_1, \ldots, A_n$, an elementary step consists in selecting an atom A_k from the query, taking a renaming $H \leftarrow B_1, \ldots, B_m$ of a clause from P, computing a most general unifier θ of A_k and H, and producing a new query $Q_{i+1} = (A_1, \ldots, A_{k-1}, B_1, \ldots, B_m, A_{k+1}, \ldots A_n)\theta$. It is required that the variables in the renamed clause have not occurred earlier in the computation. By repeating such steps we obtain an *SLD-derivation*, consisting of a sequence of queries, a sequence of substitutions, and a sequence of clauses. The atom to be replaced in each step is chosen by a *selection function*. A derivation whose last query is empty is *successful*. If Q is the initial query of a successful derivation, and $\theta_1, \ldots, \theta_l$ is its sequence of substitutions, then $Q\theta_1 \cdots \theta_l$ is the *computed answer* for P and Q.

SLD-resolution is sound; each computed answer is a logical consequence of the program. It is also complete; if $P \models Q\theta$ (an instance $Q\theta$ of a query Q is a logical consequence of P) then there exists a computed answer $Q\sigma$ for P and Q, such that $Q\theta$ is an instance of $Q\sigma$. Moreover, such a computed answer exists for any selection function.

The derivations for a given goal and a given selection rule may be represented as an *SLD-tree* (or search tree). Its root is the initial query. Each child of a node Q' is derived in one step from Q'. There is one child of Q' for each clause of P, whose (renamed) head unifies with the selected atom of Q'. Computing answers for a given program P and query Q can be seen as traversing the SLD-tree for P and Q.

SLD-resolution is the basis of most implementations of logic programming, in particular the programming language Prolog (see [3,56]). SLD-resolution is goal driven and works top-down. Bottom-up operational semantics exist too, they are useful for Datalog programs.

2.3 Logic Programming and Non-monotonic Reasoning

The need for non-monotonic reasoning. An important feature of FOL is that the reasoning is **monotonic**. This means that adding new axioms to a theory does not change the conclusions already drawn from the theory: If $T \subseteq T'$ and $T \models F$ then $T' \models F$. Also the fragment of LP introduced above (positive programs) is monotonic. Such feature is sometimes considered as a restriction.

Example 1 (Database). It is natural to represent a relational database as a logic program consisting of facts; one predicate symbol for each database relation, one fact for each tuple. For instance a database relation *salary*, registering salaries of three persons, can be represented as

$$salary(brown, 3000). \qquad salary(smith, 4000). \qquad salary(adams, 5000).$$

This set of three atoms can be treated as a logic program, or a FO theory. $salary(brown, 3000)$ follows from it, and $salary(brown, 2500)$ does not. However it does not follow from the set that $salary(brown, 2500)$ is false. (More formally, $\neg salary(brown, 2500)$ is not a logical consequence of the theory.) To be able to conclude $\neg salary(brown, 2500)$, we need a stronger FO theory, for instance

$$salary(X, Y) \leftrightarrow$$
$$X = brown, Y = 3000 \lor X = smith, Y = 4000 \lor X = adams, Y = 5000$$

in FOL augmented with equality.

Non-monotonic reasoning – positive programs. Now we discuss adding non-monotonic reasoning to logic programming with positive programs. A natural idea is the **closed world assumption** (CWA) [59]. It allows to conclude $\neg A$ when A cannot be concluded. More precisely, for a positive logic program P and a ground atom A, if $P \not\models A$ then $\neg A$ is concluded under CWA. For instance, for a ground atom A and a program P consisting of unary clauses (like the one in the example above) $\neg A$ is concluded whenever A is not an instance of a unary clause in P.

It is convenient to characterize CWA in terms of the least Herbrand models of programs. Given a program P, the complement of its least Herbrand model H_P is the set of those ground atoms whose negations are concluded by CWA.

CWA has an unpleasant computational property. In a general case, the check for $P \models A$ is undecidable. This means that there does not exist a terminating algorithm which, given P and A, checks whether $P \models A$. It is however semi-decidable: there exists such an algorithm, which terminates whenever $P \models A$, but may loop forever otherwise. Thus it is not even semi-decidable whether $\neg A$ follows from P by CWA. Fortunately, the problem is decidable and can be computed efficiently for positive Datalog programs.

From the practice of Prolog programming a competitive notion emerged: **negation as failure** (NAFF, a more precise name is *negation as finite failure*) [8]. Informally, $\neg A$ is concluded under NAFF when it can be found by a finite search that $P \not\models A$. More precisely, when (for some selection rule) the SLD-tree for A and P is finite and does not contain a successful derivation. Such a tree is called a **finitely failed** tree. NAFF is semi-decidable.

Example 2. From a program $P = \{\, p \leftarrow p \,\}$, we conclude $\neg p$ under CWA. On the other hand, nothing about p can be concluded under NAFF. The search tree for p w.r.t. P is infinite, it consists of a single infinite derivation.

For a program which is a (finite) set of unary clauses, NAFF and CWA are equivalent.

We introduced NAFF as an operational concept. It has however a logical explanation, which is provided by converting the program P to a stronger FO theory, the **completion** of P (denoted $comp(P)$), proposed by Clark [8]. We skip the definition and present only examples; the reader is referred to e.g. [4,56].

The idea is to replace the implications in P by equivalences. For the transformation to make sense, all the clauses with the same predicate in their heads are to be first grouped into a single implication. If no clause of the program begins with a predicate symbol q then the completion contains a formula $q(X_1, \ldots, X_n) \leftrightarrow$ *false*. Equality axioms (called *Clark equality theory, CET*) are added to the completion. They ensure, among others, that any two distinct ground terms denote different values. When presenting program completions we will skip CET; $comp(P) \setminus CET$ will be denoted by $comp_0(P)$.

The completion $comp(P)$ is indeed a stronger theory than P: whenever $P \models A$, for an atom A, then $comp(P) \models A$.

Example 3. In most of our examples we assume that the alphabet of predicate symbols is the set of predicate symbols occurring in the program.

The completion of the program from Ex. 1 consists of CET and the equivalence formula $salary(X, Y) \leftrightarrow \ldots$ given in the example.

For $P_1 = \{\, p \leftarrow p \,\}$ we have $comp_0(P_1) = \{\, p \leftrightarrow p \,\}$, and for $P_1' = \{\, p \leftarrow q \,\}$ we have $comp_0(P_1') = \{\, p \leftrightarrow q;\ q \leftrightarrow false \,\}$. Notice that $comp(P_1) \not\models \neg p$, and $comp(P_1') \models \neg q, \neg p$.

Example 4. Consider a program P_2:

$$e(a,b). \qquad e(a,c). \qquad c(X,Z) \leftarrow e(X,Z).$$
$$e(b,a). \qquad e(c,d). \qquad c(X,Z) \leftarrow e(X,Y), c(Y,Z).$$

attempting to define the transitive closure of a given relation. (Predicate e describes edges in a graph, and c describes pairs of connected nodes.) The completion of P_2 contains two equivalences

$$e(X,Y) \leftrightarrow X = a, Y = b \vee X = b, Y = a \vee X = a, Y = c \vee X = c, Y = d.$$
$$c(X,Z) \leftrightarrow e(X,Z) \vee \exists Y : e(X,Y), c(Y,Z).$$

It follows that $c(b,d)$, $\neg c(d,b)$ and $\neg c(c,b)$ are logical consequences of $comp(P)$; on the other hand, $\neg c(a,f)$ is not. (Let us assume, that f is a constant in our alphabet.) This is because there exists a model of $comp(P)$ in which $c(a, f)$ and $c(b, f)$ are true.

This corresponds to NAFF. Under the Prolog selection rule, the search trees for $c(d,b)$ and $c(c,b)$ are finitely failed; but the tree for $c(a,f)$ is infinite under any selection rule.

Under CWA the program correctly describes the transitive closure of e. This is because for any ground x, y it holds that $P_2 \models c(x,y)$ iff there is a path from x to y in the graph defined by e.

Example 5 (Inequality). Consider a program $P_= = \{ eq(X, X) \}$. We have $P_= \models eq(t_1, t_2)$ iff t_1, t_2 are the same term. The least Herbrand model is $H_{P_=} = \{ eq(t, t) \mid t$ is a ground term $\}$. For ground terms t_1, t_2, literal $\neg eq(t_1, t_2)$ is concluded under CWA (equivalently, under NAFF) iff t_1, t_2 are distinct.[2] We have $comp_0(P_=) = \{ eq(X, Y) \leftrightarrow X = Y \}$.

NAFF is sound and complete w.r.t. the completion semantics: There exists a finitely failed tree for P and Q iff $comp(P) \models \neg Q$, for any positive logic program P and query Q.

To summarize, there exist two main semantics of obtaining negative conclusions from positive logic programs: CWA and the completion semantics. The former is more often used in knowledge representation, the latter in Prolog programming.

3   Logic Programs with Negation

It is often required that programs are able to refer to the negative conclusions obtained by CWA of NAFF. This leads to negative literals in clause bodies, i.e. to general programs. Generalizing CWA and completion semantics to general programs is a difficult task. It has been a subject of intensive research. As a result, three main solutions emerged: 3-valued completion semantics, and two

---

[2] For non-ground t_1, t_2, literal $\neg eq(t_1, t_2)$ is concluded (under both CWA and NAFF) iff t_1, t_2 do not have a common instance.

generalizations of CWA: the well-founded semantics and stable model semantics. The first one seems not so important from the point of view of knowledge representation and Semantic Web applications; we include it in order to make our overview more comprehensive.

3.1   3-valued Completion Semantics (3CS)

We begin this section with the operational semantics. It may be claimed that a generalization of NAFF to programs with negation led the generalization of the corresponding declarative semantics – the completion semantics. In my opinion, the main purpose of the generalized completion semantics was to describe the underlying operational semantics.

NAFF can be generalized to general programs. The resulting operational semantics is called SLDNF-resolution. The main idea is applying NAFF recursively. Instead of one SLD-tree, a tree of SLNDF-trees is built. When a negative literal $\neg A$ is selected in a tree node Q then a tree for query A is built. If the tree produces A (or a renaming of A) as an answer then $\neg A$ is found to be false, the node with $\neg A$ has no children. If the tree produces answers, but none of them is a renaming of A, then Q is said to be *floundered* and has no children. The intuition is that $\neg A$ may be true for some values of its variables, but NAFF is unable to find these values. If the tree for A does not produce answers, has no floundered nodes and is finite then we say that A *finitely fails*. In this case $\neg A$ is concluded, and node Q has one child, which is Q with $\neg A$ removed.

Often a less general version of SLDNF-resolution is considered. One treats as floundered each node with a non-ground $\neg A$ selected. For formal definitions and examples, see e.g. [56,4]. The implementation of negation in the programming language Prolog is close to SLDNF-resolution. However the programmer has to take care that floundering is avoided (for instance by assuring that the selected negative literals are always ground, or by reporting floundering as a run-time error).

It turns out that SLNDF-resolution is sound w.r.t. the completion semantics, adapted from positive programs in a straightforward way. However such semantics is not feasible. Speaking informally, it implies too much.

Example 6 (Completion of general programs). Applying the notion of completion to some general programs leads to undesired results.

Consider a program P, in which predicate p does not occur. Program $P_1 = P \cup \{ p \leftarrow \neg p \}$ has an inconsistent completion; $comp(P_1) \models A$ and $comp(P_1) \models \neg A$, for each atom A. So adding such a clause makes the completion semantics of the whole program useless.

The completion of $\{ p \leftarrow \neg p, \neg q; \; q \leftarrow q \}$ is $\{ p \leftrightarrow \neg p, \neg q; \; q \leftrightarrow q \}$. It implies that q is true (and p is false). Under any generalization of NAFF the program will not produce such result; the only derivation for q is infinite (and employs only the clause $q \leftarrow q$). Removing the "meaningless" rule $q \leftarrow q$ results in a program, whose completion $\{ p \leftrightarrow \neg p, \neg q; \; q \leftrightarrow false \}$ is inconsistent. Replacing the rule by another "meaningless" rule $p \leftarrow p$ results in a program, whose completion $\{ p \leftrightarrow (\neg p, \neg q \vee p); \; q \leftrightarrow false \}$ implies p and $\neg q$ [66].

The "right" declarative semantics turns out to be that proposed by Kunen: program completion in a certain 3-valued logic [43]. (For the details the reader is referred to [4] or [14]. Basically, the notion of completion is the same as for positive programs, but it is interpreted in a nonstandard logic.) A main intuition is that the third logical value *undefined* corresponds to inability of determining whether a query is true or false. We write $T \models_3 F$ if a formula F is a 3-valued logical consequence of a theory T. So F is concluded from P under 3-valued completion semantics iff $comp(P) \models_3 F$.

For some programs, the 3-valued and 2-valued semantics are equivalent, see [44,4] for a syntactic criterion. Dealing with non-standard 3-valued logic may be inconvenient; equivalent characterizations in the standard 2-valued logic are possible, see [63,19] and references therein. They are based on introducing another completion $comp'(P)$, such that $comp(P) \models_3 L$ iff $comp'(P) \models L$ (for any literal L).

Example 7. Consider a program WIN [34,66]:

$$w(X) \leftarrow m(X,Y), \neg w(Y). \qquad m(a,b). \qquad \ldots$$

(the remaining unary clauses for m are skipped). The program describes a game. Predicate m defines the graph of the game. A position X is winning ($w(X)$ is true) if there exists a move to a losing position Y ($w(Y)$ is false).

Its completion semantics expressed in 2-valued logic (following [19], with an obvious simplification) is given by a theory $comp^*(WIN)$ consisting of equivalences

$$w(X) \leftrightarrow \exists Y : m(X,Y), \neg w'(Y) \qquad\qquad m(X,Y) \leftrightarrow X = a, Y = b \vee \ldots$$
$$w'(X) \leftrightarrow \exists Y : m(X,Y), \neg w(Y)$$

and the equality theory CET (whose role is to state that distinct constants have distinct values). The predicate w was replaced by two ones. An intuitive explanation is that w, w' have distinct logical values (in 2-valued logic) in the cases where originally w has the third logical value *undefined*. A position c is winning iff $comp^*(WIN) \models w(c)$. It is a losing position iff $comp^*(WIN) \models \neg w(c)$. Otherwise it is neither losing nor winning. For instance, a position from which no move is possible is losing; if the graph consists of a single cycle (e.g. $m(a,b)$; $m(b,c)$; $m(c,a)$) then all its positions are neither losing nor winning.

Example 8 (Inequality 2). Consider a program $P'_=$:

$$eq(X,X). \qquad\qquad neq(X,Y) \leftarrow \neg eq(X,Y).$$

(which is an extension of the program $P_=$ from Ex. 5). We state without proof that the completion of $P'_=$ under 3-valued logic is equivalent to that under 2-valued logic. We have $comp_0(P'_=) = \{ eq(X,Y) \leftrightarrow X = Y; neq(X,Y) \leftrightarrow \neg eq(X,Y) \}$.

A literal $eq(t_1, t_2)$ (or $\neg eq(t_1, t_2)$) is a consequence of $P'_=$ (under the completion semantics) iff it is a consequence of $P_=$ (under NAFF or CWA). An atom

$neq(t_1, t_2)$ (respectively a literal $\neg neq(t_1, t_2)$) is a consequence of $P'_=$ (under the completion semantics) iff $\neg eq(t_1, t_2)$ is (respectively $eq(t_1, t_2)$ is).

We also mention that for $P'_=$ the completion semantics is equivalent to the two semantics introduced further on, the well-founded and the stable model semantics.

The completion semantics is usually not considered in the context of knowledge representation. For an exception, see [40] and a reference therein.

3.2 The Well-Founded Semantics (WFS)

In the current and the next section we present two main generalizations of CWA to programs with negation. Both are expressed by means of models of programs.

It will be convenient to consider ground instances of programs; $ground(P)$ is the set of all ground instances of the clauses of the program P. Notice that $ground(P)$ depends on the considered alphabet $\mathcal{F}$ of function symbols; thus it should be made clear what $\mathcal{F}$ is. Often one requires that $\mathcal{F}$ are the function symbols occurring in P; we do not impose this restriction.

We first show that CWA is inapplicable to general programs. Consider a program $P = \{ p \leftarrow \neg q \}$. We have $P \not\models p$ and $P \not\models q$. CWA would allow concluding $\neg p$ and $\neg q$. However these results are incompatible with the program: $P \cup \{ \neg p, \neg q \}$ is unsatisfiable (as $P \cup \{ \neg q \} \models p$). From the other point of view: CWA is characterized by the least Herbrand models of programs; but for a general program the least Herbrand model may not exist.

Now we present the intuition behind the well-founded semantics (WFS). Our presentation follows [4]. For a general program P, there are certain positive and negative conclusions that we believe should be derived. For instance, if a ground atom A is a unary clause in $ground(P)$ then A should be derived. If A is not the head of any clause of $ground(P)$ then $\neg A$ should be derived. This idea can be generalized. Some ground literals follow from the program regardless of the semantics of negative literals in clause bodies: Some atoms can be derived from the positive part of $ground(P)$ (i.e. the clauses not containing negative literals). Some atoms cannot be derived, even if all the negative literals in the clause bodies are removed; if A is such atom, we should derive $\neg A$, by CWA. In this way a set I of ground literals can be concluded from the program.

Now I can be used to fix the value of some literals in $ground(P)$, and to produce more conclusions. The process is repeated until reaching a fixed point. The obtained set of literals is considered to be the semantics of P (and is called the well-founded model of P).

It can be shown that the sets of literals obtained in the consecutive steps satisfy the requirements of the following definition.

Definition 1. A 3-valued interpretation is a set I of ground literals such that if $A \in I$ then $\neg A \notin I$, for any atom $A \in \mathcal{HB}$. A ground literal L is true in I if $L \in I$, it is false if $\neg L \in I$ (when L is a negative literal $\neg B$ then by $\neg L$ we mean the atom B), otherwise it has the third logical value *unknown* (or *undefined*).

A 3-valued interpretation I is called **total** if $A \in I$ or $\neg A \in I$ for each ground atom A. A total I can be understood as a 2-valued interpretation $I \cap \mathcal{HB}$ represented as a 3-valued one. For a 2-valued interpretation J, we define the corresponding total 3-valued one: $3(J) = J \cup \{\neg A \mid A \notin J, \ A \in \mathcal{HB}\}$.

Before a formal definition of the well-founded semantics we present some examples.

Example 9. Consider a program $\{p \leftarrow p; \ q \leftarrow \neg p; \ r \leftarrow q, \neg r\}$. Regardless of the values of the negative literals in the program, p cannot be derived. Thus we derive $\neg p$. From this, in the second iteration, we derive q. In the next iteration, neither r nor $\neg r$ can be derived. Thus the well-founded model of the program is $\{\neg p, q\}$.

Example 10 ([18]). Let us come back to the program from Ex. 7:

$$w(X) \leftarrow m(X, Y), \neg w(Y)$$

$m(b, a)$	$m(c, d)$	$m(d, e)$	$d \to e$
$m(a, b)$	$m(c, f)$	$m(e, f)$	$\uparrow \quad \downarrow$
$m(a, c)$			$b \leftrightarrow a \to c \to f$

(The diagram illustrates the graph described by predicate m.) Its well-founded model contains $\neg w(f), w(e), w(c), \neg w(d)$, but neither $w(a)$, nor $\neg w(a)$, and neither $w(b)$, nor $\neg w(b)$. Notice that the well-founded semantics of this program coincides with its 3-valued completion semantics.

Example 11. Consider a program P:

$$p(a) \leftarrow \neg p(X)$$
$$p(X) \leftarrow p(X)$$
$$p(b) \leftarrow \neg p(b)$$

and assume that the alphabet of constants is $\{a, b, c\}$. Let gP_2 be the (three element) set of ground instances of the second clause of P. The positive part of $ground(P)$ is gP_2, no atom is its consequence. $ground(P)$ with negative literals removed is $gP_2 \cup \{p(a)\leftarrow; \ p(b)\leftarrow\}$. From this program, by CWA, we derive $\neg p(c)$. The obtained interpretation is $I_1 = \{\neg p(c)\}$.

Second iteration: Literal $\neg p(c) \in I_1$ occurs in the body of $p(a) \leftarrow \neg p(c) \in ground(P)$, it is removed from the clause, as it is known to be true. Now the positive part of the program is $gP_2 \cup \{p(a)\leftarrow\}$, and $p(a)$ is its consequence. The program with negative literals removed is as previously. We obtain $I_2 = I_1 \cup \{p(a)\}$.

In the third iteration, as $p(a)$ is known to be true, $p(a) \leftarrow \neg p(a)$ is removed from the program, and $p(a)$ is removed from the body of $p(a) \leftarrow p(a)$. (Also, $\neg p(c)$ is removed from a clause as previously). The positive part of the program is now $gP_2 \setminus \{p(a) \leftarrow p(a)\} \cup \{p(a)\leftarrow\}$. It produces the same consequences, i.e. $\{p(a)\}$, as that in the previous iteration. The program without negative literals is now $gP_2 \setminus \{p(a) \leftarrow p(a)\} \cup \{p(a)\leftarrow; \ p(b)\leftarrow\}$. Applying CWA to it results in a single literal $\neg p(c)$, as in the previous iterations. Thus a fixpoint $I_3 = I_2 = \{\neg p(c), p(a)\}$ is reached; I_2 is the well-founded model of P. Hence in the well-founded semantics of P, $p(a)$ is true, $p(c)$ is false, and $p(b)$ is unknown.

The well-founded semantics was introduced by van Gelder, Ross and Schlipf [66]. There exist several equivalent definitions, here we follow [4]. (For the others, see e.g. [4] and the references in [4,66], [17, p. 51], [5, p. 87].)

Definition 2 (The well-founded semantics)
Let P be a general program. The positive part of P, denoted by P^+, is P without all the clauses containing a negative literal. By P^- we denote P with all the negative literals deleted. (Notice that P^+ and P^- are positive programs; we will use their least Herbrand models $\mathcal{M}_{P^+} \subseteq \mathcal{M}_{P^-}$.) Let

$$\mathcal{I}_3(P) = \mathcal{M}_{P^+} \cup \{ \neg A \mid A \notin \mathcal{M}_{P^-} \}.$$

Let P be a program and I a 3-valued interpretation. $P|I$ denotes the program obtained from $ground(P)$ by deleting all clauses that contain a body literal that is false in I, and deleting all body literals that are true in I. An iteration step is described by

$$\Phi_P(I) = \mathcal{I}_3(P|I).$$

The operator Φ_P is monotonic, hence it has its least fixed point, which is called the **well-founded model** of P, and denoted by $WF(P)$.

If a formula F is true in $WF(P)$ then we say that F is *entailed* by the WFS (the well-founded semantics) of P (or F is a *consequence* of P under WFS).

The fixed point of Φ_P can be obtained by iterating $I_{n+1} = \Phi_P(I_n)$, starting from $I_0 = \emptyset$. A transfinite sequence of steps may be necessary in a general case; for a limit ordinal β we have $I_\beta = \bigcup_{\alpha < \beta} I_\alpha$. If $ground(P)$ is finite then a finite sequence is sufficient.

Example 12. Assume that the alphabet contains a constant 0 and a unary function symbol s. More than ω iterations are needed for the program:

$$p \leftarrow odd(X), \neg odd(X)$$
$$odd(s(X)) \leftarrow \neg odd(X)$$

We have $I_{2n} = \{ odd(s^1(0)), \ldots, odd(s^{2n-1}(0)), \neg odd(0), \ldots, \neg odd(s^{2n-2}(0)) \}$ for any natural number n, and then $I_\omega = \{ odd(s^1(0)), odd(s^3(0)), \ldots, \neg odd(0), \neg odd(s^2(0)), \ldots \}$. The well founded model is $I_{\omega+1} = I_\omega \cup \{\neg p\}$. Notice that the model is total.

The reader is encouraged to check that CWA for positive programs is a special case of WFS. As CWA is not decidable (not even semi-decidable), neither is WFS. It is however decidable for Datalog programs.

An abstract operational semantics for WFS can be obtained by appropriately generalizing SLDNF-resolution. The result is called SLS-resolution [57,62,4]. The main difference is that a failed tree is not required to be finite. For an implementation, Prolog is augmented by loop checking mechanisms [7,6], which are able to discover some infinite derivations ("positive loops") and some infinite sequences of trees ("negative loops"). As WFS is undecidable, such system is unable to compute all the program answers required by the semantics (and it loops in some cases). An implementation of WFS is the XSB system [64].

3.3   Stable Model Semantics (SMS)

Here we present the second major generalization of CWA for general programs, proposed by Gelfond and Lifschitz [34]. It employs 2-valued Herbrand interpretations. Remember that such interpretations are (represented as) sets of ground atoms; an atom A is true in an interpretation I if $A \in I$, and false otherwise.

The main idea can be explained as follows. An interpretation M is used to valuate the negative literals in the clause bodies of $ground(P)$. This boils down to removing the clauses containing $\neg A$, where $A \in M$, and removing each literal $\neg A$, where $A \notin M$. The resulting program is positive. If its least Herbrand model is again M then M is a *stable model* of P. Intuitively, M was chosen as a hypothesis, and the hypothesis has been confirmed.

Definition 3 (Stable model [34]**).** Let M be a 2-valued interpretation and P^M be the program obtained from $ground(P)$ by deleting

- each rule with a negative literal $\neg A$ in its body, such that $A \in M$, and
- all negative literals in the bodies of the remaining rules (the literals are true w.r.t. M).

P^M is positive. If the least Herbrand model of P^M is M (i.e. $\mathcal{M}_{P^M} = M$) then M is called a **stable model** of P.

We say that that a formula F is *entailed* by the SMS (stable model semantics) of a program P, if F is true in every stable model of P. Sometimes one considers F to be entailed by P, when F is true in some stable model of P. The former way of reasoning is called *cautious*, or *skeptical*, reasoning, the latter *brave*, or *credulous* reasoning. Often neither brave, nor cautious consequences are of main interest. Instead, distinct stable models of a program represent distinct solutions to the problem represented by the program.

Example 13. Program $\{\, p \leftarrow \neg q \,\}$ has one stable model $\{p\}$ (as q is false in $\mathcal{M}_{\{p \leftarrow \neg q\}^M}$ for any M, and if q is false in M then p is true in $\mathcal{M}_{\{p \leftarrow \neg q\}^M}$). Program $\{\, p \leftarrow \neg p \,\}$ does not have a stable model.

Program $\{\, p \leftarrow \neg q; \ q \leftarrow \neg p \,\}$ has two stable models $\{p\}$ and $\{q\}$. If the two rules are the only rules for p and q in a program P then in any stable model of P exactly one of atoms p, q is true.

Example 14. Program WIN from Ex. 10 has two stable models. In one of them the atoms with the predicate symbol w are $\{\, w(a), w(c), w(e) \,\}$, and in the other $\{\, w(b), w(c), w(e) \,\}$.

If we replace the definition of m by $\{\, m(a,b); \ m(b,c); \ m(c,a) \,\}$ then the resulting program has no stable models.

Notice that for this program the stable model semantics is incompatible with its intended meaning, explained in Ex. 7, and provided by the well-founded semantics (and the 3-valued completion semantics). An informal explanation of WIN under SMS is that we are looking for a division of the graph nodes into winning and losing ones, such that from each winning node there is an arc to a losing one.

Example 15. The program

$$odd(s(X)) \leftarrow \neg odd(X)$$

has a unique stable model $\{\, odd(s^{2i+1}(0)) \mid i \geq 0 \,\}$, under an assumption that s and 0 are the only function symbols. (An explanation follows from three facts: $odd(0) \notin \mathcal{M}_{PM}$ for any M. If $odd(s^i(0)) \notin M$ then $odd(s^{i+1}(0)) \in \mathcal{M}_{PM}$. If $odd(s^i(0)) \in M$ then $odd(s^{i+1}(0)) \notin \mathcal{M}_{PM}$.)

The program

$$o(X) \leftarrow \neg o(s(X))$$

has two stable models: $\{\, o(s^{2i+1}(0)) \mid i \geq 0 \,\}$, and $\{\, o(s^{2i}(0)) \mid i \geq 0 \,\}$. (If $o(s^i(0)) \notin \mathcal{M}_{PM}$ then $o(s^i(0)) \leftarrow \notin P^M$, thus $o(s^{i+1}(0)) \in M$. If $o(s^i(0)) \in \mathcal{M}_{PM}$ then $o(s^i(0)) \leftarrow \in P^M$, thus $o(s^{i+1}(0)) \notin M$. And both $o(0) \in M$ or $o(0) \notin M$ are possible).

Example 16 (Integrity constraints). An integrity constraint is a requirement that a given conjunction of literals is false. We present a common programming technique to express integrity constraints under SMS. Consider a program $P = P_0 \cup \{\, f \leftarrow \overline{L}, \neg f \,\}$, where $\overline{L}$ is a sequence of literals, and predicate f does not occur in P_0. Predicate f is false in any stable model M of P. (As if f is true in M then it is false in $\mathcal{M}_{PM}$.) Thus $\overline{L}$ must be false (as $\neg f$ is true in M, hence if $\overline{L}$ is true in M then f is true in $\mathcal{M}_{PM}$, contradiction.)

Example 17 (Graph colouring, SMS [25]). Assume that the graph is represented by predicates predicate $edge/2$ and $node/1$ (for each node n of the graph, the program contains the fact $node(n)$, and for each edge (n_1, n_2) the fact $edge(n_1, n_2)$). We express 3-colouring of the graph by three unary predicates r, g, b; $r(n)$ is true when r is the colour assigned to n, and so on. These clauses ensure that each node has exactly one colour:

$$b(X) \leftarrow node(X), \neg r(X), \neg g(X)$$
$$r(X) \leftarrow node(X), \neg b(X), \neg g(X)$$
$$g(X) \leftarrow node(X), \neg r(X), \neg b(X)$$

These clauses ensure that the neighbour nodes do not have the same colour:

$$f \leftarrow \neg f, b(X), b(Y), edge(X, Y)$$
$$f \leftarrow \neg f, r(X), r(Y), edge(X, Y)$$
$$f \leftarrow \neg f, g(X), g(Y), edge(X, Y)$$

This is the whole program. Each stable model of it represents a correct colouring of the graph.

Example 18 (Dependency on the alphabet). The program $\{p(a); q \leftarrow \neg p(X); f \leftarrow \neg f, \neg q \}$ [11] does not have a stable model when the alphabet of function symbols is $\mathcal{F} = \{a\}$. It has a stable model $\{p(a), q\}$ when the alphabet is $\mathcal{F}' = \{a, b\}$ (or any other proper superset of $\mathcal{F}$).

Answer Set Programming. Often the class of general programs with SMS is generalized, by adding explicit integrity constraints, the so-called classical negation, and disjunction in rule heads (disjunctive rules). Logic programming with SMS generalized for such programs is usually called *answer set programming* (ASP). An explicit integrity constraint is a clause $\leftarrow \overline{L}$, its meaning is that $\overline{L}$ is false in each stable model. (So it is equivalent to clause $f \leftarrow \neg f, \overline{L}$, as discussed in the example above.) We explain briefly the classical negation below; we omit discussion of disjunctive rules[3] [24]. Integrity constraints and classical negation can be seen as syntactic sugar, however introducing disjunctive rules does change the power of the semantics.

The need for classical negation [35] (called also strong or explicit negation) is related to the fact that in SMS all predicates are subject to (a generalization of) CWA, and the logic is two-valued. Whenever A is not found to be true, $\neg A$ is concluded. In some cases such behaviour is not required.

Example 19 (Classical negation). This example is attributed to John McCarthy in [35]. Assume that atom *train* is derived when a train is approaching. To reason whether a bus may cross the railway track we may use the rule *cross* $\leftarrow \neg train$. However the rule is wrong when it is possible that the information used to derive *train* is incomplete. In such case CWA should not be applied. A more adequate rule is *cross* $\leftarrow no\_train$, where *no_train* is true when a train is surely not approaching. It is possible that neither *train* nor *no_train* is true, but never both of them are true.

It is sometimes needed that, speaking informally, a certain atom p may take three logical values instead of two. We can encode the three values by introducing a new predicate *no_p*, and introducing rules for p and *no_p* that appropriately describe the problem at hand. We should assure that p and *no_p* are never both true, for instance by a constraint $\leftarrow p, no\_p$. It is convenient to add *no_* as a new syntactic operator (the classical negation),[4] and modify the definition of a stable model accordingly. (Such stable models are usually called *answer sets*.) Adding classical negation does not make the SMS more powerful. Programs with classical negation can be transformed to programs without it, as the discussion above suggests.

Example 20 (Classical negation 2). In Ex. 14 we showed that the program WIN under SMS does not have the expected meaning, which it has under the WFS, see Ex. 10 (and under 3-valued completion semantics, Ex. 7). Here are clauses which, under SMS, play the role of the clause $w(X) \leftarrow m(X,Y), \neg w(Y)$ under WFS. Notice that predicate *lose* plays the role of the classical negation of *win*.

$$win(X, N) \leftarrow m(X, Y), \; succ(M, N), \; lose(Y, M).$$
$$lose(Y, N) \leftarrow \neg\, escape(Y, N).$$
$$escape(X, N) \leftarrow m(X, Y), \; \neg\, win(Y, N).$$

---

[3] A definition of a stable model of a disjunctive program is obtained from Df. 3 by replacing "the least Herbrand model of P^M" by "a minimal Herbrand model of P^M."

[4] Sometimes symbol $\neg$ is used to denote classical negation, and symbol *not* for the non-monotonic negation, denoted here by $\neg$.

The program WINSM consists of these three clauses, clauses defining the relation $m/2$, and clauses defining the successor relation $succ/2$ on natural numbers. For graphs not exceeding a given size, a finite subset of this relation is sufficient. It can be defined by unary clauses of the form $succ(0,1); succ(1,2); \ldots$ (where $1, 2, \ldots$ are constants from $\mathcal{F}$).

A position x is a winning (respectively losing) one, if $win(x,n)$ ($lose(x,n)$) is in the (unique) stable model of WINSM, for some natural number n. The informal meaning of $win(x,n)$ is that x can win in $\leq n$ double moves; $lose(x,n)$ means that x loses in $\leq n$ double moves, and $escape(x,n)$ means that up to n double moves x can avoid losing. (In particular, $win(x,0)$ is false, and thus $escape(x,0)$ is true and $lose(x,n)$ is false, for any x.)

We mention that the WFS and SMS of WINSM coincide, $WF(WINSM) \cap \mathcal{HB}$ is the unique stable model of WINSM.

For a more comprehensive introduction, further examples, explanation and discussion see e.g. [52,25,33].

Implementations. Algorithms for evaluating SMS differ substantially from those for the completion semantics and WFS (which stem from SLD-resolution). Most of them impose restrictions on the class of programs dealt with. The main restriction is excluding non-constant function symbols. (Hence the example SMS programs above containing such symbols are of little practical usage.) Some implementations impose syntactic restrictions on rules, for instance Datalog safeness (each variable occurring in a rule has to occur in a positive literal in the rule body). Further explaining of evaluation methods for SMS is outside of the scope of this paper. There exist practical and efficient implementations. The most important ones seem to be Smodels[5] [65], DLV[6] [46], CLASP[7] [32], and CMODELS[8] [36]. See e.g. [25] for further references. There exist interesting application examples, see e.g. [55,45], `http://www.kr.tuwien.ac.at/research/projects/WASP/showcase.html`.

3.4 Comparison

We first briefly compare the 3-valued completion semantics with the well-founded semantics. The main part of this section is comparison of the well-founded semantics and stable model semantics. We present some formal results, and show some examples showing that each of them is more suitable for certain tasks.

Theorem 4. *Let P be a program and L a ground literal. If $comp(P) \models_3 L$ then $L \in WF(P)$. So whatever is entailed by the 3-valued completion semantics, is entailed by the WFS.*

The theorem follows immediately from the fact that $WF(P)$ is a 3-valued model of $comp(P)$ [66].

---

[5] `http://www.tcs.hut.fi/Software/smodels/`

[6] `http://www.dbai.tuwien.ac.at/proj/dlv/`

[7] `http://www.cs.uni-potsdam.de/clasp/`

[8] `http://www.cs.utexas.edu/users/tag/cmodels/`

Informally, the main difference between the completion semantics and WFS is similar to that of NAFF and CWA. From a programmer's point of view it may be called "dealing with positive recursion". Such recursion happens (in the propositional case) when the program contains a rule $p \leftarrow \ldots, p, \ldots$ or, more generally, rules $p_0 \leftarrow \ldots, p_1, \ldots;\ \ p_1 \leftarrow \ldots, p_2, \ldots;\ \ \cdots;\ p_n \leftarrow \ldots, p_0, \ldots.$ We may say that a predicate p_i depends on itself, and the dependency does not involve negation. Under WFS, $\neg p_i$ is concluded (for $i = 1, \ldots, n$). Neither $\neg p_i$, nor p_i is concluded under the completion semantics.

WFS and SMS, formal properties. We present here some formal results concerning the relations between the well-founded and stable model semantics.

Theorem 5 ([66]). Consider the (3-valued) well-founded model $WF(P)$ of a program P. If J is a (2-valued) stable model of P then $WF(P) \subseteq 3(J)$ (cf. Df. 1).

In other words, if a literal L is true in the well-founded model of P then it is true in each stable model of P.

Theorem 6 ([66]). *If the well-founded model of P is total then the corresponding 2-valued interpretation $WF(P) \cap \mathcal{HB}$ is the unique stable model of P.*

Example 21 ([4]). The converse of the Theorem does not hold. The program $\{ p \leftarrow \neg p;\ p \leftarrow \neg q;\ q \leftarrow \neg p \}$ has a unique stable model $\{p\}$, but its well-founded model is $\emptyset$.

The WFS and SMS coincide on an important class of programs.

Definition 7 (Local stratification [4]). A *local stratification* is a function σ from $\mathcal{HB}$ to the countable ordinals. (In many cases it is sufficient to consider natural numbers as the values of σ.) For a negative literal $\neg A$, we define $\sigma(\neg A) = \sigma(A) + 1$.

A ground clause $H \leftarrow L_1, \ldots, L_n$ is *locally stratified w.r.t.* a local stratification σ if $\sigma(H) \geq \sigma(L_i)$ for $i = 1, \ldots, n$.

A program P is *locally stratified w.r.t.* σ if each clause of $ground(P)$ is. A program P is **locally stratified** if it is locally stratified w.r.t. some local stratification.

Given a local stratification σ, we sometimes call $\sigma(A)$ the level of A. Local stratification σ divides $ground(P)$ into strata, according to the levels of clause heads. So the stratum numbered j is $P_j = \{ A \leftarrow \overline{L} \in ground(P) \mid \sigma(A) = j \}$, and $P = \bigcup_j P_j$. If a negative literal $\neg A$ occurs in P_j, then A does not occur in any clause head in any P_k with $k \geq j$. If an atom A occurs in P_j, then A does not occur in any clause head in any P_k with $k > j$. The first non-empty stratum is a positive program. The intuition behind stratification is that it the meaning of atoms of higher level depends on the meaning of those of lower level; the level is strictly lower if negation is involved.

The well-founded semantics of a locally stratified program P can be obtained following the ordering of the strata. At each step a single stratum is considered, not the whole program as in Df. 2:

$$J_0 = \emptyset,$$
$$Q_i = P_i | J_i,$$
$$J_{i+1} = \mathcal{M}_{Q_i} \cup \{\, \neg A \mid \sigma(A) = i, \ A \notin \mathcal{M}_{Q_i} \,\} \cup J_i,$$

for $i \geq 0$, and

$$WF(P) = \bigcup_i J_i.$$

The union is over all i such that $\sigma(A) = i$ for some A. Each program Q_i is positive, and all its atoms are of level i. The atoms from $\mathcal{M}_{Q_i}$ are true in the 3-valued interpretation J_{i+1}, hence true in the well-founded model $WF(P)$ of P. All the remaining atoms of level i are false in J_{i+1}, and in $WF(P)$. Thus each atom is either true or false in some J_k. We obtain

Theorem 8. If P is locally stratified then its well-founded model $WF(P)$ is total (see [4]). Thus the corresponding two-valued interpretation $WF(P) \cap \mathcal{HB}$ is its unique stable model.

Example 22. The program from Ex. 12 is locally stratified, under a σ such that $\sigma(odd(s^i(0))) = i$ and $\sigma(p) = \omega$. The first program from Ex. 15 is locally stratified under the same local stratification.

The second program $\{\, o(X) \leftarrow \neg o(s(X)) \,\}$ from Ex. 15 is not locally stratified. (As $\sigma(o(s^i(0))) > \sigma(o(s^{i+1}(0)))$ is required, for each $i = 0, 1, \ldots$; but there does not exist and infinite decreasing sequence of ordinals.)

Program WIN from Ex. 10 is not locally stratified, as $w(a) \leftarrow m(a, a), \neg w(a)$ is in $ground(WIN)$, and Df. 7 requires $\sigma(w(a)) \geq \sigma(\neg w(a)) = \sigma(w(a)) + 1$, contradiction.

Local stratification generalizes a simpler, standard notion of stratification, where the numbers are assigned to predicate symbols, not to atoms. The case of program WIN from the last example suggests further generalization. If $m(a, a)$ is false in $WF(WIN)$ then the clause $w(a) \leftarrow m(a, a), \neg w(a)$, which violates the conditions of Df. 7, is in a sense irrelevant. Under a generalized stratification, program WIN with an acyclic game graph can be considered stratified; its well-founded model is total. The reader is referred to [4] for further discussion and references.

Pragmatics. A main difference between WFS and SMS may be informally formulated as follows. SMS is based on making hypotheses, and checking whether a hypothesis is compatible with the program. (A hypothesis is an interpretation M, the check consists of constructing P^M and finding whether its least Herbrand model is M again.) On the other hand, in WFS no hypotheses are made, the well founded model contains only those conclusions which, intuitively, are unavoidable.

A related view is presented in [12]: Programs under WFS can be seen as inductive definitions, but under SMS a program "is viewed not as a set of definitions but as a set of rules expressing constraints on the problem domain".

In contrast to SMS, in the WFS the meaning of p is influenced only by the predicates on which p syntactically depends [13]. (This property is called *relevance*.) More precisely: Let us say that a predicate p directly syntactically depends on a predicate q in a program P, if P contains a clause for p (i.e. a clause with p in its head) with q in its body. Let us define recursively that p *syntactically depends* on q in P if p directly depends on q or p directly depends on some r, and r depends on q. Now consider the set $P' \subseteq P$ of the clauses for p and for those predicates on which p depends in P. Then the meaning of p in the well-founded semantics of P and of P' is the same. Formally, $p(\bar{t}) \in WF(P)$ iff $p(\bar{t}) \in WF(P')$, and $\neg p(\bar{t}) \in WF(P)$ iff $\neg p(\bar{t}) \in WF(P')$.

The SMS does not have this property. For instance consider an integrity constraint $C = f \leftarrow p, \neg f$. Predicate p does not depend on f, but removing C from P can change the meaning of p. (Take $P' = \{p \leftarrow \neg q; \ q \leftarrow \neg p\}$ and $P = \{C\} \cup P'$; atom p is true in some stable models of P', but false in each stable model of P.)

A usual style of using SMS is constructing a program so that each stable model corresponds to a solution of the problem dealt with. In contrast, in WFS to each solution the corresponds an answer to the program.

Example 23. Ex. 17 presents a graph colouring program under SMS. Here we solve the same problem under WFS. The part of the program defining the graph is the same. An atom $c(l)$ is in the $WF(P)$ iff l is a correct colouring of the graph, represented as a list of pairs $p(n, c)$, where n is a node of the graph, and c is its colour.[9]

$$c(\,[p(N, C)]\,) \leftarrow colour(C), node(N)$$
$$c(\,[p(N, C)|L]\,) \leftarrow c(L), colour(C), node(N), \neg bad(N, C, L)$$
$$bad(N, C, L) \leftarrow edge(N, M), member(p(M, C), L)$$
$$bad(N, C, L) \leftarrow edge(M, N), member(p(M, C), L)$$
$$bad(N, \_, L) \leftarrow member(p(N, \_), L)$$

An atom of the form $bad(n, c, l)$ expresses the fact that node n with colour c is incompatible with the partial colouring l (because l assigns c to a node neighbouring n, or l assigns a colour to n.) The program should be completed by defining auxiliary predicates:

$$colour(r) \qquad member(X, [X|\_])$$
$$colour(g) \qquad member(X, [\_|L]) \leftarrow member(X, L).$$
$$colour(b)$$

---

[9] The list notation of Prolog (see e.g. [56]) is used. $[h|t]$ denotes a list with the head h and the tail t. $[h]$ stands for one element list, where the element is h. We use so-called anonymous variables of Prolog: Each occurrence of $\_$ stands for a distinct variable (whose name is irrelevant).

Notice that in the SMS version a solution is represented as a stable model (more precisely, by the restriction of a stable model to predicates r, g, b). In the WFS, to each solution there corresponds an answer to the program; the solution is represented by a data structure – a list of pairs. In this program, r, g, b are constants.

We presented some examples of solving the same problem under WFS and SMS. The reader is encouraged to compare them. For some problems WFS is more natural and convenient than SMS, for some problems the situation is opposite. For the WIN / WINSM program (Ex. 10, 20) the WFS is more suitable – the WFS program is definitely simpler. (Also, creating the SMS program from Ex. 20 was not so obvious for the author.) The comparison looks different for the graph colouring problem (Ex. 17, 23). In this case the SMS program seems simpler and easier to understand. Some tasks, like integrity constraints are not expressible in WFS in a natural way. SMS, and in particular its generalization Answer Set Programming, seems more popular and has strong supporters. An important practical difference is that implementations of SMS (and of answer set programming) are usually restricted to Datalog (non-constant function symbols are forbidden). This makes it impossible to use data structures, which in Logic Programming are constructed by means of function symbols with arguments.

4   Combining Rules and Description Logics

In the context of the Semantic Web there are established formalisms related to reasoning with ontologies, namely RDF Schema and the web ontology languages OWL and OWL 2 [67]. The mathematical basis of OWL and OWL 2 are Description Logics (DL). Another reasoning approach considered for the Semantic Web is inference based on rule sets. This approach is the main motivation of activities of W3C RIF Working Group.[10] It is generally expected that both ontology reasoning and rule based reasoning will be used, and – moreover – both approaches will be combined. An important difference is that reasoning with ontologies is monotonic, while rules are a convenient formalism for non-monotonic reasoning, as discussed above. Thus one may expect that integrating monotonic and non-monotonic reasoning should be an important aspect of combining rules and description logics. This aspect is the focus of this chapter.

It is desirable that integration of rules and ontologies could employ the existing reasoning tools for LP and DL, instead of building new tools from scratch. So we are mainly interested in those formalisms for which such re-using of tools is possible. We omit approaches called *homogeneous* (in the terminology of [51,2]). A homogeneous approach (to combining an LP language and a DL) provides a formalism which includes both original ones as special cases. There is no distinction between the rule predicates and DL predicates. Usually, existing reasoners for LP or DL cannot be employed by reasoning algorithms for the new formalism.

We mention a few examples of homogeneous integration. In SWRL [41], the union of an ontology and of a set of rules is considered as a single FO theory, and

---

[10] http://www.w3.org/2005/rules/wiki/RIF_Working_Group

the semantics is based on the standard logical consequence of FOL. Thus non-monotonic reasoning is not supported. MKNF$^+$ knowledge bases [54] embed rules and ontologies in a common formalism based on the logic of minimal knowledge and negation as failure (MKNF) by Lifschitz [49]. The semantics is related to the SMS. Hybrid MKNF knowledge bases of [42,1] are also based on MKNF, but with a semantics related to the well-founded semantics. The unifying formalism applied to rules and ontologies in [11] is quantified equilibrium logic.

Alternatively, homogeneous integration is achieved by some appropriate restriction of the LP or DL part. For instance, Description Logic Programs (DLP) [37], consider a specific restricted description logic which can be encoded by a logic program without negation, treated as a first order theory. (So the reasoning is monotonic.)

For an overview of approaches to integration of rules and description logics, the reader is referred to [51,20,39,2,26].

4.1   Loose and Tight Coupling

We discuss *heterogeneous* [51] (or *hybrid*) integration of rules and ontologies. This group of approaches preserves the distinction between the rule predicates and ontology predicates. The integration is achieved by permitting ontology predicates in the rules. In most cases, reasoning in an integrated formalism can be based on reasoning tools for the underlying LP and DL formalisms.

So we expect to deal with a **hybrid program** (P, T) consisting of a set of rules (a logic program) P and a first order *external theory* T. In most cases T is a DL ontology. We divide the alphabet $\mathcal{P}$ of predicate symbols into $\mathcal{P}_P$ and $\mathcal{P}_T$ called, respectively, the sets of *rule predicates* and *external predicates*. The predicates from $\mathcal{P}_P$ cannot appear in T. The predicates from $\mathcal{P}_T$ cannot appear in the rule heads of P. Thus, informally, the external theory is independent from P and defines the predicates from $\mathcal{P}_T$. On the other hand, P defines the predicates from $\mathcal{P}_P$, possibly using the predicates defined by T. The atoms (literals) with the predicates from $\mathcal{P}_T$ ($\mathcal{P}_P$) will be called, respectively, *external* (*rule*) atoms (literals).

The alphabet of function symbols $\mathcal{F}$ is the same for the rule component P and the first order component T; $\mathcal{F}$ may contain symbols which do not occur in (P, T). We assume that for a given program $\mathcal{F}$ is fixed. The ground instance $ground(P)$ of P depends on the choice of $\mathcal{F}$. We permit function symbols of non-zero arity, but in most approaches referred to below, $\mathcal{F}$ contains only constants.

The semantics of T is that of FOL. For the semantics of P two main choices are to be done, in principle the choices are independent. The first one is the semantics of negation in P. Most commonly the stable model semantics is employed. The WFS is the second alternative. The other choice is the semantics of external literals in the rules. We now discuss two possibilities, loose coupling and tight coupling (of P and T) [51,20]. (These two notions are similar to interaction based on entailment and based on single models of [10].)

In **loose coupling**, the semantics of P is defined in terms of interpretations of $\mathcal{P}_P$ and the logical consequence of T. A ground external literal L is treated, in a sense, like a procedure call. Its logical value is true if it is a logical consequence of T; otherwise its value is false. So the logical value of a rule body in an interpretation is defined in terms of logical consequence of the external theory. The two usually disjoint levels of semantics: truth in an interpretation and logical consequence, are mixed here.

In the case of **tight coupling**, the semantics of P is defined in terms of interpretations of $\mathcal{P}_P \cup \mathcal{P}_T$. The logical value of an external literal in an interpretation is defined in a standard way. Before giving a formal definition, let us present an example.

Example 24 (Reasoning by cases [51,20]). Assume that the external theory T classifies courses as project courses and lecture courses:

$$Project(X) \vee Lecture(X) \leftrightarrow Course(X)$$
$$Lecture(cs05)$$
$$Project(cs21)$$
$$Course(cs14)$$
$$\ldots$$

The rules P:

$$student(X) \leftarrow enrolled(X,Y), Lecture(Y)$$
$$student(X) \leftarrow enrolled(X,Y), Project(Y)$$
$$enrolled(ann, cs14)$$
$$\ldots$$

define a student as a person enrolled in a lecture or a project. (So *student, enrolled* are rule predicates, and the remaining predicates are external.) We have $T \models Course(cs14)$, but $T \not\models Lecture(cs14)$, $T \not\models Project(cs14)$. Thus in loose coupling both $Lecture(cs14)$ and $Project(cs14)$ are false and $student(ann)$ cannot be derived.

In tight coupling, in every interpretation $Lecture(cs14)$ is true or $Project(cs14)$ is true. Thus $student(ann)$ can be derived, as in each interpretation $student(ann)$ is true, due to one of the rules

$$student(ann) \leftarrow enrolled(ann, cs14), Lecture(cs14)$$
$$student(ann) \leftarrow enrolled(ann, cs14), Project(cs14)$$

from $ground(P)$.

An important technical detail should be made clear. The symbol $\neg$ denotes two kinds of negation. In an external literal, $\neg$ denotes the standard negation of FOL. In a rule literal, $\neg$ denotes the non-monotonic negation (of the chosen semantics for rules). To improve readability it may be reasonable to distinguish syntactically the two kinds of negation, for instance by introducing a symbol *not* for the non-monotonic negation.[11]

Definition 9 (Loose coupling). Consider a hybrid program (P,T). Let $P/_{\models}T$ be the program obtained from $ground(P)$ by deleting

---

[11] Such syntax gives some new possibilities, e.g. we can write $p \leftarrow not \neg q$ instead of $nonq \leftarrow \neg q; \ p \leftarrow \neg nonq$.

- each rule containing an external literal L which is not a logical consequence of T (i.e. $T \not\models L$), and
- all external literals in the bodies of the remaining rules ($T \models L$ for each such literal L).

A stable model of $P/\models T$ is called a *stable model* of (P,T) *under loose coupling.* The well-founded model of $P/\models T$ is called the *well-founded model of (P,T) under loose coupling.*

Definition 10 (Tight coupling). Consider a hybrid program (P,T), and a (2-valued) model M_0 of T. Let P/M_0 be the program obtained from $ground(P)$ by deleting

- each rule containing an external literal L which is not true in M_0, (i.e. $M_0 \not\models \neg L$), and
- each external literal in the bodies of the remaining rules ($M_0 \models L$ for each such literal L).

The well-founded model $WF(P/M_0)$ of P/M_0 is called the *well-founded model* of (P,T) *based on* M_0. A formula F over $\mathcal{P}_P$ holds (or *is true*) in the *well-founded semantics under tight coupling* (WFST) if F is true in each well-founded model of (P,T).

A stable model of P/M_0 is called a *stable model* of (P,T) *based on* M_0. A formula F over $\mathcal{P}_P$ is *entailed* by the *stable model semantics under tight coupling* (SMST) if F is true in each stable model of (P,T).

Notice that M_0 is a model over an arbitrary domain, while the stable and well-founded models are over the Herbrand universe.

Example 25. [18,23] Here we present a variant of the game from Ex. 7, with some restrictions imposed on the moves. Imagine that the positions represent geographical locations described by an ontology T. Assume that from the ontology it follows that any location in Finland is a location in Europe, $T \models Fi(X) \to E(X)$. Let the moves of the game be defined as in the program P:

$$w(X) \leftarrow m(X,Y), \neg w(Y)$$

$$
\begin{array}{ll}
m(b,a) & m(c,f) \leftarrow \neg Fi(f) \\
m(a,b) & m(e,f) \leftarrow E(f) \\
m(a,c) & \\
m(c,d) & \\
m(d,e) &
\end{array}
$$

$$
\begin{array}{ccc}
d & \longrightarrow & e \\
\uparrow & & \downarrow {\scriptstyle E(f)} \\
b \leftrightarrow a \to c & \underset{\neg Fi(f)}{\longrightarrow} & f
\end{array}
$$

A move from c to f is allowed if f is in not in Finland, and a move from e to f if f is in Europe. As previously, position f is losing. If $E(f)$ is not a logical consequence of T then we cannot conclude that e is a winning positions. (There exists a model of T in which f is not in Europe, and the move from e to f is impossible.) However when $E(f)$ is true then e is winning, d is losing, and c is winning. Otherwise, $E(f)$ is false, hence $Fi(f)$ is false, the move from c to f is possible, and c is winning. Thus c is always a winning position.

This intuition is reflected by WFST (the well-founded semantics under tight coupling). If $M_0 \models E(f)$ then $WF(P/M_0) \models w(e), \neg w(d), w(c)$. Otherwise $M_0 \models Fi(f)$, and $WF(P/M_0) \models w(c)$. Thus $w(c)$ holds in the WFST of (P, T). On the other hand, $w(e)$ is true in some well-founded models of (P, T), and is false in others; hence neither $w(e)$, nor $\neg w(e)$ holds in the WFST.

Assume that $T \not\models E(f)$ and $T \not\models \neg Fi(f)$. Then in the well-founded model under loose coupling, we have both $\neg m(c, f)$ and $\neg m(e, f)$, and thus $\neg w(e)$, $w(d)$, and $\neg w(c)$. The semantics is not as intended.

Example 26. Assume that insurance should be arranged for X if it does not follow from the external theory T that X is insured. This can be expressed under loose coupling (but not under tight coupling) by rules P:

$$insured(X) \leftarrow Insured(X).$$
$$insurance\_needed(X) \leftarrow \neg insured(X).$$

Here *Insured* is an external predicate (and the remaining predicate symbols are rule predicates). The program is stratified, so the WFS and SMS are equivalent. As required, under loose coupling $insurance\_needed(a)$ is derived iff $T \not\models Insured(a)$ (for any constant a). It is derived under tight coupling iff $Insured(a)$ is false in all the models of T.

The last three examples show that the expressive powers of loose and tight coupling are complementary. There are tasks that can be expressed by one of them, but not by the other. In loose coupling, reasoning by cases is impossible. On the other hand, under loose coupling the rule component P can query whether a literal L is, or is not, a logical consequence of T. (Hence it can query if $T \cup \{L\}$ is consistent.) This is impossible under tight coupling. So we conclude that it is useful to introduce a more powerful kind of hybrid programs, which could use both kinds of coupling.

We now present a hybrid program which uses non-constant function symbols to construct data structures.

Example 27 (A non Datalog program [21]). Here an additional requirement to the game from Ex. 25 is added. Each node can be visited at most once. The list of forbidden nodes is kept in the second argument of predicate $win/2$. The semantics is WFST.

$w(X) \leftarrow win(X, [])$
$win(X, History) \leftarrow move(X, Y, History), \neg win(Y, [X|History])$
$move(A, B, History) \leftarrow m(A, B), \neg member(B, History)$

$m(b, a)$	$m(c, f) \leftarrow \neg Fi(f)$
$m(a, b)$	$m(e, f) \leftarrow E(f)$
$m(a, c)$	
$m(c, d)$	$member(X, [X|T])$
$m(d, e)$	$member(X, [H|T]) \leftarrow member(X, T)$

In this variant of the game, for the given graph, both a and b become winning positions (and for the remaining nodes the game behaves as in Ex. 25).

Equality. The semantics of hybrid programs integrates a semantics of LP, based on Herbrand interpretations, and the standard semantics of FOL, based on arbitrary interpretations. In the former, in contrast to the latter, each element of the domain is a value of some ground term and the values of distinct ground terms are distinct. The difference between two kinds of interpretations may lead to somehow strange results. The external theory T may imply that some ground terms a, b are equal. But it is possible that in a stable or well-founded model of (P, T) an atom $p(a)$ is true but $p(b)$ is false (for instance consider $P = \{p(a)\}$).

We accept this feature of the semantics, as we prefer to keep the standard semantics for T. The user can construct programs in such a way that whenever (P, T) entails $p(a_1, \ldots, a_n)$, and $T \models a_1 = b_1, \ldots, a_n = b_n$ then (P, T) entails $p(b_1, \ldots, b_n)$ (and the analogical property holds for negative literals). Informally, whenever T implies $a = b$ then P should treat a and b in the same way. We will call such programs *congruent*. For instance, if for each constant a occurring in P, T does not imply that a is equal to some other constant then (P, T) is congruent. See [18, Section 3.3] for further discussion of this issue.

Example 28. [18] Consider the hybrid program from Ex. 25 and assume that $\mathcal{F}$ contains a constant g. Under WFST the program entails $w(c)$ and $\neg w(g)$. Assume T implies $c = g$. Then the program is not congruent.

We can make it congruent by modifying P. Assume that T does not imply $c_1 = c_2$ for any other pair of constants $\{c_1, c_2\} \neq \{c, g\}$. Then it is sufficient to add to P new rules $m(a, g); m(g, d); m(g, f) \leftarrow \neg Fi(f)$. (We replaced c by g in the rules of P.) Now – informally – c and g are treated by the obtained hybrid program (P', T) in the same way, and – formally – (P', T) is congruent.

The problem is known from constraint logic programming (CLP) [53], where terms equal in the constraint domain may be non-unifiable, and thus treated as having distinct values. (For instance in most versions of CLP over arithmetic constraints, terms 4 and $2 + 2$ are not unifiable, but denote the same number and are treated as equal by arithmetic constraint predicates.) Apparently programmers do not find this confusing, and can easily cope with the problem using some ad hoc programming techniques.

An alternative solution to the problem of equality is to modify the FOL semantics of T, by imposing a restriction on interpretations that distinct ground terms denote distinct values. (This however disallows most ways of using equality in T.) If $\mathcal{F}$ contains only constants then this requirement is called *unique names assumption*, UNA. Many approaches to integrating rules with DL restrict FOL by UNA, e.g. CARIN [48], $\mathcal{AL}$-log [15], f-hybrid knowledge bases [31], r-hybrid KBs and $\mathcal{DL}$+*log* [60,61]. Yet another treatment of equality is proposed in [54]. Roughly speaking, the interpretations have to satisfy UNA. However, the equality of T is interpreted as an equivalence relation (more precisely, as a congruence w.r.t. the interpretation of predicates); the values of terms are equivalent when their values under the standard FOL semantics are equal.

Decidability. Let us now discuss decidability of hybrid reasoning. Assume that the Herbrand universe is finite (i.e. $\mathcal{F}$ is a finite set of constants). For decidability of (P, T) under loose coupling it is sufficient that it is decidable whether $T \models L$ for any external literal L occurring in $ground(P)$. For decidability under WFST and SMST it is sufficient that for a set S of ground external literals satisfiability of $T \cup S$ is decidable.[12]

Decidability of hybrid reasoning contrasts with the fact that checking $P \cup T \models A$ is undecidable already for rules without negation and for most description logics of interest [48]. However in the latter case $\models$ deals with arbitrary interpretation domains, while the semantics of Df. 10, 9 is based on Herbrand interpretations of P.

Loose coupling approaches. An important representative are dl-programs [29], and their descendants (cf. Section 4.2). The considered external theories are the description logics underlying OWL Lite and OWL. Dl-programs generalize the ideas of Df. 9 in two ways. First, the rules are disjunctive rules (under the answer set semantics). Then, the ways of querying the external theories are richer than those of Df. 9; we discuss this in Section 4.2. There also exists a variant of dl-programs for (non-disjunctive) rules with WFS [28].

Tight coupling approaches. We now outline some existing tight coupling approaches.

CARIN [48] and $\mathcal{AL}$-log [15] are classical systems combining positive Datalog programs with certain description logics. They use the FOL semantics, but with the unique name assumption, for both rules and DL. The main difference from the semantics of Df. 10 is that the considered interpretations of the rules are not restricted to Herbrand interpretations. (As FOL – the semantics of DL – is applied here to the rules, this approach may also be understood as homogeneous integration of rules and DL.)

HD-rules [18,23] is a formalism with tight coupling and the well-founded semantics. Its semantics follows that of Df. 10. In contrast to most of the formalisms outlined here, it permits non-constant function symbols.

The following approaches employ disjunctive Datalog rules with answer set semantics, in this way generalizing SMST of Df. 10. Moreover they permit external predicates in rule heads.

The formalism of r-hybrid KBs was introduced in [60] (KB stands for "knowledge base"), and generalized in $\mathcal{DL}+log$ [61]. In this approach, $\mathcal{F}$ is a countably infinite set of constants, and the interpretation domain is fixed so that there is a one-to-one correspondence between constants and domain elements. This requirement on the domain is called *standard names assumption*. To achieve decidability, syntactic restrictions (so-called weak DL-safeness) are imposed on

---

[12] Let S be the set of external literals in $ground(P)$. The set of programs of the form P/I, where I is an interpretation for T, is finite. Let $S_I = \{ L \in S \mid I \models S \} \cup \{ \neg L \in S \mid I \models \neg S \}$. If $S_{I_1} = S_{I_2}$ then $P/I_1 = P/I_2$. $T \cup S_I$ is satisfiable iff there exists a model M_0 of T such that $P/M_0 = P/I$. Thus all the programs P/M_0 satisfying the conditions of Df. 10 can be enumerated by checking satisfiability of $T \cup S_I$ for each possible S_I. Hence all the well-founded (stable) models of (P, T) can be computed. Another sufficient condition for decidability of (a generalization of) WFST is given by [18, Th. 5.1].

rules in r-hybrid KBs. Moreover, negative external literals are not allowed. Weak DL-safeness means that each variable of a rule has to occur in a positive rule literal in the body, unless the variable occurs only in external literals in the body.

The formalism of f-hybrid knowledge bases (fKBs) [31] uses the description logic $\mathcal{SHOQ}$ and open answer set programming. The latter means that, instead of a fixed alphabet $\mathcal{F}$, stable models over an arbitrary, maybe infinite, set of constants are considered. To ensure decidability, the form of rules is restricted. The restricted class of disjunctive programs, called forest logic programs (FoLPs), is sufficient to encode $\mathcal{SHOQ}$. So the considered hybrid programs can be translated to logic programs. (Thus the formalism may be also seen as a homogeneous approach, because a single formalism covers both the rules and the DL part of hybrid programs.)

Tightly coupled dl-programs [50] combine DL with disjunctive rules, without syntactic restrictions. This approach differs from the other ones in considering, roughly speaking, minimal models of the external theories. For instance, p is entailed by a program (P, T) with $P = \{p \leftarrow \neg q(b)\}$ and $T = \{q(a)\}$ (where $P_P = \{p\}$, $P_T = \{q\}$), despite $\neg q(b)$ is not entailed by T in the standard semantics of DL.

Reasoning. Query evaluation (or finding stable models) under loose coupling can be implemented in a rather straightforward way. One can use an implementation of LP with negation (for WFS or SMS). Whenever the evaluation algorithm needs the logical value of a ground external atom, this value is obtained from a reasoner for the external theory. An important example of such implementation is a system NLP-DL for dl-programs [29]. Using DLV system [47] as an LP engine, it is able to deal with SMS and WFS (restricted to Datalog). The external theory is a DL, and the employed reasoner is Racer [38].

For tight coupling such obvious and general methods of reasoning do not exist. The work on r-hybrid KBs and $\mathcal{DL}+log$ [60,61] proposes algorithms based, roughly speaking, on enumeration of all Herbrand interpretations over an restricted Herbrand base (for the external predicates and constants occurring in the rules). Given a program (P, T), for each such interpretation M the theory T is queried (with a certain query related to M) and existence of a stable model is checked (of a program similar to P/M). The number of models to be enumerated is exponential in the number of constants in P. We are not aware of any implementations.

For f-hybrid knowledge bases [31] an implementation method is proposed based on translation to a restricted class of disjunctive logic programs.

An important approach is based on ideas from the operational semantics of constraint logic programming (CLP). It was introduced for the case of rules without negation in the early work on $\mathcal{AL}$-log [15]. For WFS with tight coupling a general algorithm was proposed for HD-rules in [18,23,21,22]. It stems from SLS-resolution [58], which is a usual operational semantics for the WFS, and from its generalization to so-called constructive negation [16,17]. Computations are top-down (query driven). A tree of trees is constructed, similarly as in SLS- and SLDNF-resolution. Moreover, queries to the external theory T are obtained from the trees. Constructing the trees is related to P, and may be seen as the upper layer of the computation, Querying T may be seen as the subsidiary layer of the computation.

The operational semantics of $\mathcal{AL}$-log [15] and of HD-rules [18,23] coincide for rules without negation. We illustrate the semantics on a brief example.

Example 29. Consider the program (P, T) from Ex. 24 and a query $student(X)$. A tree is built employing the rules for $student$ from P.

$$student(X)$$

$$\diagup \qquad \diagdown$$

$$enrolled(X, Y), Lecture(Y) \qquad\qquad enrolled(X, Y), Project(Y)$$

$$| \qquad\qquad\qquad\qquad\qquad |$$

$$X = ann,\ Y = cs14,\ Lecture(cs14) \qquad X = ann,\ Y = cs14,\ Project(cs14)$$

From the tree it follows that $Lecture(cs14) \vee Project(cs14)$ implies $student(ann)$. A query is issued whether $T \models Lecture(cs14) \vee Project(cs14)$. The answer is yes, and $X = ann$ is returned as an answer to the initial query.

The reader is referred to [18] or [21] for definitions and further examples. A corresponding implementation [21] employs XSB Prolog [64] as an LP engine, and Pellet [9] as an DL reasoner. The reasoner is treated as a black box and, in principle, any DL reasoner can be used. Non-constant function symbols are allowed. The implementation is at a prototype stage, and is available at `http://www.ida.liu.se/hswrl/`.

4.2   Generalizations

Most implementations of SMS allow disjunctive rules. So a natural generalization for hybrid programs with SMS is to permit such rules. This is done by some approaches outlined above.

According to Df. 9, 10, the rules in a hybrid program (P, T) refer to the external theory by means of external literals. It may be useful to replace literals by a more general class of formulae over $\mathcal{P}_T$. In HD-rules [21,18] it is an arbitrary class of formulae, closed under certain logical operations and satisfying a decidability condition (if C is such a formula with its free variables replaced by constants then is should be decidable whether $T \models C$). In dl-programs [29], a (possibly negated) concept inclusion query may appear in a rule body. Such a query $C \sqsubseteq D$ is true if $C(X) \rightarrow D(X)$ is a logical consequence of the external theory T. Another generalization of external literals in dl-programs is explained below; it is related to modifying T. Further generalizations are present in the descendants of dl-programs: HEX-programs [30] and cq-programs [27] (see also [20] for an introduction).

Influencing external theory by rules. The hybrid programs, as defined above, are asymmetric. In such a program (P, T), external predicates may occur in rules P, but rule predicates cannot occur in the external theory T. So, rule predicates are defined in terms of both P and T, but external predicates are, apparently, defined in terms of T alone. Sometimes it is however desirable that the rules can influence the external theory. We show three ways of achieving this.

Such possibility already exists in SMST. For some models M_0 of T, a stable model of P/M_0 may not exist. In this way the rules P influence T by excluding some of its models.

Example 30. Assume that $T = \emptyset$ and $\mathcal{P}_T = \{q\}$ (q is the only external predicate). Let $P = \{f \leftarrow \neg f, \neg q; \ p \leftarrow q\}$. Consider SMST. There are two ground programs to consider: $P/M_0 = \{f \leftarrow \neg f\}$, and $P/M_1 = \{p\}$ (where $M_0 = \emptyset$, $M_1 = \{q\}$ are Herbrand models of T). Program P/M_0 does not have a stable model, but P/M_1 does. So, we may say that P makes the external predicate q to be true. The unique stable model of P/M_1 is $\{p\}$, and this is the only stable model of (P, T).

A possibly more natural way of influencing the external theory by the rules is allowing external predicates in the rule heads. Such feature is present in r-hybrid KBs and $\mathcal{DL}+log$ [60,61], f-hybrid KBs [31], and in tightly coupled dl-programs of [50]. As they are SMST approaches, we conjecture that each program can be transformed to an equivalent one without external predicates in the rule heads, using a technique suggested by Ex. 30. Instead of a rule $q \leftarrow \overline{L}$, where q is an external predicate, we may use $\leftarrow \neg q, \overline{L}$ (equivalently $f \leftarrow \neg f, \neg q, \overline{L}$).

These two ways of influencing the external theory seem to make sense only with tight coupling and SMS. In contrast, the next approach works with loose coupling. In dl-programs [29], external literals in rule bodies are generalized to *dl-atoms*. Roughly speaking, a dl-atom consists of an external literal L, and an expression λ describing a modification of T. The modification is adding some facts to T; the facts are described by λ in terms of (the meaning of) the rule predicates. The literal L is to be evaluated in the modified theory $T \cup T_\lambda$, where T_λ is the set of facts specified by λ. (See [29] for definitions, examples and further explanations.)

5   Conclusion

We discussed certain approaches to integrating rules with external theories (more formally – logic programs with FOL theories). The approaches are heterogeneous: the semantics of hybrid programs is constructed out the FOL semantics of the external theories and the non-monotonic semantics of logic programs, WFS or SMS. Implementation methods are possible that employ existing reasoners for FOL and for logic programs under WFS or SMS. For other ways of integrating rules and ontologies, and for further references, the reader is referred to other overview papers, for instance [51,20,39,2,26].

Heterogeneous integration of rules with external theories provides an array of possibilities. There is a choice of the semantics of negation in logic programs, between the well-founded semantics (WFS), and the stable model semantics (SMS) or its generalization – answer set semantics for disjunctive logic programs. In

Section 3, we compared these semantics. There is also a choice between loose and tight coupling between the rule part and the FOL part; this choice leads to substantially different results. Loose coupling seems less elegant, as it mixes the semantic levels of logical consequence and of truth in an interpretation. It fails when reasoning by cases is needed, as in Ex. 24. However it has a useful ability of querying whether a literal is a logical consequence of the external theory. Such ability is sometimes necessary, as Ex. 26 shows. Also, implementing loose coupling is simpler than tight coupling, and at least one mature implementation exist [29]. An implementation of tight coupling with WFS [21,18] is at a prototype stage.

Intuitively, the information flow between external theories and rules is unidirectional: the rule predicates are defined by the rules and the external theory, but the external predicates only by the external theory. We however showed that SMS under tight coupling is an exception. In this case the rules can change the meaning of external predicates (cf. Ex. 30). We discussed some generalizations of the basic paradigm of hybrid programs, including those which make it possible to influence external theories by rules.

As loose and tight coupling differ semantically, we expect that applications will require both of them, to express different things (as illustrated by the aforementioned examples). Thus it seems necessary to construct a hybrid language that combines both kinds of coupling.

Acknowledgements. Comments of Jan Małuszyński and three anonymous referees helped to improve this paper. The author is solely responsible for the remaining drawbacks.

References

1. Alferes, J.J., Knorr, M., Swift, T.: Queries to hybrid MKNF knowledge bases through oracular tabling. In: Bernstein, A., Karger, D.R., Heath, T., Feigenbaum, L., Maynard, D., Motta, E., Thirunarayan, K. (eds.) ISWC 2009. LNCS, vol. 5823, pp. 1–16. Springer, Heidelberg (2009)
2. Antoniou, G., Damásio, C.V., Grosof, B., Horrocks, I., Kifer, M., Maluszynski, J., Patel-Schneider, P.F.: Combining rules and ontologies. A survey. FP6 NoE REWERSE, Deliverable I3-D3,
   http://rewerse.net/deliverables/m12/i3-d3.pdf
3. Apt, K.R.: From Logic Programming to Prolog. Prentice-Hall, Englewood Cliffs (1997)
4. Apt, K.R., Bol, R.N.: Logic Programming and Negation: A Survey. Journal of Logic Programming 19/20, 9–71 (1994)
5. Baral, C., Gelfond, M.: Logic Programming and Knowledge Representation. Journal of Logic Programming 19/20, 73–148 (1994)
6. Bol, R.N., Degerstedt, L.: Tabulated resolution for the well-founded semantics. J. Log. Program. 34(2), 67–109 (1998)
7. Chen, W., Warren, D.S.: Tabled evaluation with delaying for general logic programs. J. ACM 43(1), 20–74 (1996)

8. Clark, K.L.: Negation as Failure. In: Gallaire, H., Minker, J. (eds.) Logic and Data Bases, pp. 293–322. Plenum Press, New York (1978)
9. Clark & Parsia, LLC: Pellet: The open source OWL 2 reasoner, `http://www.mindswap.org/2003/pellet/index.shtml`, `http://clarkparsia.com/pellet`
10. de Bruijn, J., Eiter, T., Polleres, A., Tompits, H.: On representational issues about combinations of classical theories with nonmonotonic rules. In: Lang, J., Lin, F., Wang, J. (eds.) KSEM 2006. LNCS (LNAI), vol. 4092, pp. 1–22. Springer, Heidelberg (2006)
11. de Bruijn, J., Pearce, D., Polleres, A., Valverde, A.: A semantical framework for hybrid knowledge bases. Knowledge and Information Systems (to appear, 2010)
12. Denecker, M., Bruynooghe, M., Marek, V.W.: Logic programming revisited: Logic programs as inductive definitions. ACM Trans. Comput. Log. 2(4), 623–654 (2001)
13. Dix, J.: A Classification Theory of Semantics of Normal Logic Programs: II. Weak Properties. Fundamenta Informaticae 22, 257–288 (1995)
14. Doets, K.: From Logic to Logic Programming. MIT Press, Cambridge (1994)
15. Donini, F., Lenzerini, M., Nardi, D., Schaerf, A.: $\mathcal{AL}$-log: Integrating Datalog and description logics. Intelligent Information Systems 10(3), 227–252 (1998)
16. Drabent, W.: SLS-resolution without floundering. In: Pereira, L.M., Nerode, A. (eds.) Proc. 2nd International Workshop on Logic Programming and Non-Monotonic Reasoning, pp. 82–98. MIT Press, Cambridge (June 1993)
17. Drabent, W.: What is failure? An approach to constructive negation. Acta Informatica 32(1), 27–59 (1995)
18. Drabent, W., Maluszynski, J.: Hybrid Rules with Well-Founded Semantics. To appear in Knowledge and Information Systems (2010)
19. Drabent, W.: Completeness of SLDNF-resolution for non-floundering queries. J. Logic Programming 27(2), 89–106 (1996)
20. Drabent, W., Eiter, T., Ianni, G., Krennwallner, T., Lukasiewicz, T., Maluszynski, J.: Hybrid reasoning with rules and ontologies. In: Bry, F., Maluszynski, J. (eds.) REWERSE 2009. LNCS, vol. 5500, pp. 1–49. Springer, Heidelberg (2009)
21. Drabent, W., Henriksson, J., Maluszynski, J.: HD-rules: A hybrid system interfacing Prolog with DL-reasoners. In: Proceedings of the ICLP'07 Workshop on Applications of Logic Programming to the Web, Semantic Web and Semantic Web Services (ALPSWS 2007), CEUR Workshop Proceedings, vol. 287 (2007), `http://www.ceur-ws.org/Vol-287`
22. Drabent, W., Henriksson, J., Maluszynski, J.: Hybrid reasoning with rules and constraints under well-founded semantics. In: Marchiori, M., Pan, J.Z., de Sainte Marie, C. (eds.) RR 2007. LNCS, vol. 4524, pp. 348–357. Springer, Heidelberg (2007)
23. Drabent, W., Małuszyński, J.: Well-founded Semantics for Hybrid Rules. In: Marchiori, M., Pan, J.Z., de Sainte Marie, C. (eds.) RR 2007. LNCS, vol. 4524, pp. 1–15. Springer, Heidelberg (2007)
24. Eiter, T., Gottlob, G., Mannila, H.: Disjunctive Datalog. ACM Transactions on Database Systems 22(3), 364–418 (1997)
25. Eiter, T., Ianni, G., Krennwallner, T.: Answer set programming: A primer. In: Tessaris, S., Franconi, E., Eiter, T., Gutierrez, C., Handschuh, S., Rousset, M.C., Schmidt, R.A. (eds.) Reasoning Web. Semantic Technologies for Information Systems. LNCS, vol. 5689, pp. 40–110. Springer, Heidelberg (2009)

26. Eiter, T., Ianni, G., Krennwallner, T., Polleres, A.: Rules and ontologies for the Semantic Web. In: Baroglio, C., Bonatti, P.A., Maluszynski, J., Marchiori, M., Polleres, A., Schaffert, S. (eds.) Reasoning Web. LNCS, vol. 5224, pp. 1–53. Springer, Heidelberg (2008), Slides available at, http://rease.semanticweb.org/
27. Eiter, T., Ianni, G., Krennwallner, T., Schindlauer, R.: Exploiting conjunctive queries in description logic programs. Ann. Math. Artif. Intell. 53(1-4), 115–152 (2008) doi:10.1007/s10472-009-9111-3
28. Eiter, T., Ianni, G., Lukasiewicz, T., Schindlauer, R.: Well-founded semantics for description logic programs in the semantic web. Research Report 1843-09-01, INFSYS, Technische Universität, Wien. To appear in ACM Transactions on Computational Logic (2009)
29. Eiter, T., Ianni, G., Lukasiewicz, T., Schindlauer, R., Tompits, H.: Combining answer set programminag with description logics for the Semantic Web. Artificial Intelligence 172(12-13), 1495–1539 (2008)
30. Eiter, T., Ianni, G., Schindlauer, R., Tompits, H.: Effective Integration of Declarative Rules with External Evaluations for Semantic Web Reasoning. In: Sure, Y., Domingue, J. (eds.) ESWC 2006. LNCS, vol. 4011, pp. 273–287. Springer, Heidelberg (2006)
31. Feier, C., Heymans, S.: Hybrid reasoning with forest logic programs. In: Aroyo, L., Traverso, P., Ciravegna, F., Cimiano, P., Heath, T., Hyvönen, E., Mizoguchi, R., Oren, E., Sabou, M., Simperl, E. (eds.) ESWC 2009. LNCS, vol. 5554, pp. 338–352. Springer, Heidelberg (2009)
32. Gebser, M., Kaufmann, B., Neumann, A., Schaub, T.: clasp: A conflict-driven answer set solver. In: Baral, C., Brewka, G., Schlipf, J. (eds.) LPNMR 2007. LNCS (LNAI), vol. 4483, pp. 260–265. Springer, Heidelberg (2007)
33. Gelfond, M.: Answer sets. In: van Harmelen, F., Lifschitz, V., Porter, B. (eds.) Handbook of Knowledge Representation. Elsevier, Amsterdam (2007)
34. Gelfond, M., Lifschitz, V.: The stable model semantics for logic programming. In: Kowalski, R.A., Bowen, K. (eds.) Proceedings of the Fifth International Conference on Logic Programming, pp. 1070–1080. MIT Press, Cambridge (1988)
35. Gelfond, M., Lifschitz, V.: Classical Negation in Logic Programs and Disjunctive Databases. New Generation Computing 9, 365–385 (1991)
36. Giunchiglia, E., Lierler, Y., Maratea, M.: Answer set programming based on propositional satisfiability. J. Autom. Reasoning 36(4), 345–377 (2006)
37. Grosof, B.N., Horrocks, I., Volz, R., Decker, S.: Description logic programs: Combining logic programs with description logics. In: Proceedings of the Twelfth International World Wide Web Conference, WWW 2003, Budapest, Hungary, pp. 48–57 (2003)
38. Haarslev, V., Möller, R.: Description of the RACER system and its applications. In: DL 2001 Workshop on Description Logics, Stanford, CA (2001)
39. Hitzler, P., Parsia, B.: Ontologies and rules. In: Staab, S., Studer, R. (eds.) Handbook on Ontologies, 2nd edn., pp. 111–132. Springer, Heidelberg (2009)
40. Hölldobler, S., Ramli, C.D.P.K.: Logic programs under three-valued Lukasiewicz semantics. In: Hill, P.M., Warren, D.S. (eds.) ICLP 2009. LNCS, vol. 5649, pp. 464–478. Springer, Heidelberg (2009)

41. Horrocks, I., Patel-Schneider, P.F., Bechhofer, S., Tsarkov, D.: OWL rules: A proposal and prototype implementation. J. Web Sem. 3(1), 23–40 (2005)
42. Knorr, M., Alferes, J.J., Hitzler, P.: A coherent well-founded model for hybrid MKNF knowledge bases. In: Ghallab, M., Spyropoulos, C.D., Fakotakis, N., Avouris, N.M. (eds.) ECAI. Frontiers in Artificial Intelligence and Applications, vol. 178, pp. 99–103. IOS Press, Amsterdam (2008)
43. Kunen, K.: Negation in logic programming. Journal of Logic Programming 4(4), 289–308 (1987)
44. Kunen, K.: Signed data dependencies in logic programs. J. Log. Program. 7(3), 231–245 (1989)
45. Leone, N.: Exploiting ASP in real-world applications: Main strengths and challenges. In: Erdem, E., Lin, F., Schaub, T. (eds.) LPNMR 2009. LNCS, vol. 5753, pp. 628–630. Springer, Heidelberg (2009)
46. Leone, N., Faber, W.: The DLV project: A tour from theory and research to applications and market. In: de la Banda, M.G., Pontelli, E. (eds.) ICLP 2008. LNCS, vol. 5366, pp. 53–68. Springer, Heidelberg (2008)
47. Leone, N., Pfeifer, G., Faber, W., Eiter, T., Gottlob, G., Perri, S., Scarcello, F.: The DLV System for Knowledge Representation and Reasoning. ACM Transactions on Computational Logic 7(3), 499–562 (2006)
48. Levy, A., Rousset, M.: CARIN: A representation language combining Horn rules and description logics. Artificial Intelligence 104(1-2), 165–209 (1998)
49. Lifschitz, V.: Nonmonotonic databases and epistemic queries. In: Proceedings IJCAI-91, pp. 381–386 (1991)
50. Lukasiewicz, T.: A novel combination of answer set programming with description logics for the Semantic Web. In: Franconi, E., Kifer, M., May, W. (eds.) ESWC 2007. LNCS, vol. 4519, pp. 384–398. Springer, Heidelberg (2007)
51. Małuszyński, J.: Integration of rules and ontologies. In: Liu, L., Özsu, M.T. (eds.) Encyclopedia of Database Systems, pp. 1546–1551. Springer, US (2009)
52. Marek, V.W., Truszczyński, M.: Stable models and an alternative logic programming paradigm. In: Apt, K.R., Marek, V.W., Truszczyński, M., Warren, D.S. (eds.) The Logic Programming Paradigm – A 25-Year Perspective, pp. 375–398. Springer, Heidelberg (1999)
53. Marriott, K., Stuckey, P.J., Wallace, M.: Constraint logic programming. In: Handbook of Constraint Programming. Elsevier, Amsterdam (2006)
54. Motik, B., Rosati, R.: Reconciling description logics and rules. J. ACM 57(5) (2010)
55. Niemelä, I.: Answer set programming: A declarative approach to solving search problems. In: Fisher, M., van der Hoek, W., Konev, B., Lisitsa, A. (eds.) JELIA 2006. LNCS (LNAI), vol. 4160, pp. 15–18. Springer, Heidelberg (2006)
56. Nilsson, U., Małuszyński, J.: Logic, Programming and Prolog, 2nd edn. John Wiley and Sons, Chichester (1995), http://www.ida.liu.se/~ulfni/pp/
57. Przymusinski, T.C.: Every logic program has a natural stratification and an iterated least fixed point model. In: PODS, pp. 11–21. ACM Press, New York (1989)
58. Przymusinski, T.C.: On the Declarative and Procedural Semantics of Logic Programs. Journal of Automated Reasoning 5(2), 167–205 (1989)
59. Reiter, R.: On Closed World Data Bases. In: Gallaire, H., Minker, J. (eds.) Logic and Data Bases, pp. 55–76. Plenum Press, New York (1978)
60. Rosati, R.: On the decidability and complexity of integrating ontologies and rules. Journal of Web Semantics 3(1), 61–73 (2005)

61. Rosati, R.: $\mathcal{DL}+log$: Tight integration of Description Logics and disjunctive Datalog. In: Doherty, P., Mylopoulos, J., Welty, C.A. (eds.) KR, pp. 68–78. AAAI Press, Menlo Park (2006)
62. Ross, K.A.: A prodedural semantics for well-founded negation in logic programs. J. Log. Program. 13(1), 1–22 (1992)
63. Stärk, R.F.: From logic programs to inductive definitions. In: Hodges, W.A., et al. (eds.) Logic: From Foundations to Applications, European Logic Colloquium'93, pp. 453–481. Clarendon Press, Oxford (1996)
64. Swift, T., Warren, D.S., et al.: The XSB System. Version 3.2., Programmer's Manual, vol. 1 (2009), http://xsb.sourceforge.net
65. Syrjänen, T., Niemelä, I.: The Smodels system. In: Eiter, T., Faber, W., Truszczyński, M. (eds.) LPNMR 2001. LNCS (LNAI), vol. 2173, pp. 434–438. Springer, Heidelberg (2001)
66. Van Gelder, A., Ross, K.A., Schlipf, J.S.: The Well-Founded Semantics for General Logic Programs. Journal of the ACM 38(3), 620–650 (1991)
67. World Wide Web Consortium: OWL Web Ontology Language Current Status, http://www.w3.org/standards/techs/owl/

Model Driven Engineering with Ontology Technologies*

Steffen Staab, Tobias Walter, Gerd Gröner, and Fernando Silva Parreiras

Institute for Web Science and Technology, University of Koblenz-Landau
Universitätsstraße 1, Koblenz 56070, Germany
{staab,walter,groener,parreiras}@uni-koblenz.de

Abstract. Ontologies constitute formal models of some aspect of the world that may be used for drawing interesting logical conclusions even for large models. Software models capture relevant characteristics of a software artifact to be developed, yet, most often these software models have limited formal semantics, or the underlying (often graphical) software language varies from case to case in a way that makes it hard if not impossible to fix its semantics. In this contribution, we survey the use of ontology technologies for software modeling in order to carry over advantages from ontology technologies to the software modeling domain. It will turn out that ontology-based metamodels constitute a core means for exploiting expressive ontology reasoning in the software modeling domain while remaining flexible enough to accommodate varying needs of software modelers.

1 Introduction

Today *Model Driven Development* (MDD) plays a key role in describing and building software systems. A variety of software modeling languages may be used to develop one large software system. Each language focuses on different views and problems of the system [1]. *Model Driven Engineering* (MDE) is related to the design and specification of modelling languages, and it is based on the four-layer modelling architecture [2]. In such a modelling architecture, the M0-layer represents the real world objects. Models are defined at the M1-layer, a simplification and abstraction of the M0-layer. Models at the M1-layer are defined using concepts which are described by metamodels at the M2-layer. Each metamodel at the M2-layer determines how expressive its models can be. Analogously, metamodels are defined by using concepts described as metametamodels at the M3-layer.

Although the four-layer modelling architecture provides the basis for formally defining software modelling languages, we have identified some open challenges. Semantics of modelling languages often is not defined explicitly but hidden in modelling tools. To fix a specific formal semantics for metamodels, it should

---

* This work has been funded by the European Commission within the 7th Framework Programme project MOST no. ICT-2008-216691, http://most-project.eu.

U. Aßmann, A. Bartho, and C. Wende (Eds.): Reasoning Web 2010, LNCS 6325, pp. 62–98, 2010.

be defined precisely in the metamodel specification. The syntactic correctness of models is often analyzed implicitly using procedural checks of the modelling tools. To make well-formedness constraints more explicit, they should be defined precisely in the metamodel specification.

OWL 2, the web ontology language, is a W3C recommendation with a very comprehensive set of constructs for concept definitions [3] and allow for specifying formal models of domains. Ontologies are conceptual models, that can be described by OWL. Based on its underlying formal semantics different services are provided, which vary between satisfiability checking at the model layer, checking the consistency of instances with regard to the model, or classifying instances (finding their possible types) with regard to instance and type descriptions. Since ontology languages are described by metamodels and allow for describing structural and behavioural models, they provide the capability to combine them with software modelling languages.

In this chapter, we tackle challenges of defining both semantics and syntactic constraints, restricting the use of the abstract syntax of a modelling language, for software languages. We show how ontologies can support the definition of software modelling language semantics and provide the definition of syntactic constraints. Since OWL 2 has not been designed to act as a metamodel for defining modelling languages, we propose to build such languages in an integrated manner by bridging pure language metamodels and an OWL metamodel in order to benefit from both approaches.

This chapter is structured as follows: In Section 2, we give a short introduction of Model Driven Engineering (MDE). In Section 3, we consider ontology languages and technologies. Here we present the ontology language OWL 2. Furthermore, we consider standard ontology reasoning services and services for explanation and model repair. In Section 4 we present architectures for bridging software modelling languages and ontology technologies. Section 5 and 6 are dealing with ontology reasoning for modelling languages, where Section 5 considers structural modelling languages and Section 6 considers behavioral modelling languages. Section 7 presents the TwoUse Toolkit which implements most of the approaches presented in this paper. Section 8 gives some related work and Section 9 concludes the paper.

2   Model-Driven Engineering

The approach of model-driven software engineering (MDE) [1] suggests first to develop models describing a system in an abstract way, which is transformed in several steps into real, executable systems (e.g. source code).

Figure 1 illustrates a generic process of model driven engineering. Here a software designer starts with creating a model n, which conforms to a modelling language n. Model n describes the system by a very abstract representation.

By transforming the model n to model $n-1$, which conforms to another modelling language $n-1$, the software designer increases the platform specificity and simultaneously lowers the level of abstraction. At the end of the MDE process, source code may be generated from model 1, where the source code (e.g. Java code) conforms to some EBNF grammar.

In general MDE consists of the following two main artifacts: *Modelling languages*, which are used to describe a set of models and *model transformations*, which are used to translate models represented in one language into models represented in another language. Both, modelling languages and model transformations are introduced in the following two sections.

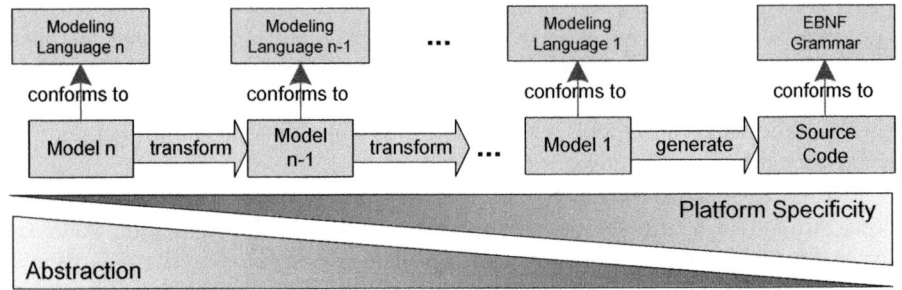

Fig. 1. Overview of MDE

2.1   Modelling Languages

The meaning of a model must be well-defined such that multiple developers can understand and work with it. If the meaning is not clear and unique there would be no possibility to define automated transformations from one abstract model to a more specific model. Furthermore, an agreed meaning of a model can be based on a common language defining precise syntax and semantics used for describing a model. In MDE models are described by modelling languages, where modelling languages themselves are described by so called metamodelling languages. A modelling language consists of an *abstract syntax*, at least one *concrete syntax* and *semantics*.

The abstract syntax of a modelling language is described by a metamodel and is designed by a *language designer*. A metamodel is a model that defines the concepts and reference for expressing a model. Semantics of the language may be defined by a natural language specification or may be captured (partially) by logics. A concrete syntax which could be of a textual or visual kind is used by a *language user* to create software models. Since metamodels are also models metamodelling languages are needed, to describe modelling languages. Here the abstract syntax is described by a metametamodel.

In the scope of graph-based modelling to create software models [4] a metamodelling language (e.g. grUML [5]) must allow for defining graph schemas, which provide types for vertices and edges, and structures them in hierarchies. In this case, each graph is an instance of its corresponding *graph schema*.

The Meta-Object Facility (MOF) is OMG's standard for defining metamodels. It provides a language for defining the abstract syntax of modelling languages.

MOF is in general a minimal set of concepts which can be used for defining other modelling languages. The version 2.0 of MOF provides two metametamodels, namely *Essential MOF* (EMOF) and *Complete MOF* (CMOF). EMOF prefers simplicity of implementation before expressiveness. CMOF instead is more expressive, but more complicated to implement [6]. EMOF mainly consists of the Basic package of the Unified Modelling Language (UML) which is part of the UML infrastructure [7]. It allows for defining classes together with properties, which are used to describe data attributes of classes and which allow for referring to other classes.

Figure 2 depicts a simplified version of the ECore metametamodel. A metamodel described by Ecore consists of Packages which can be nested and contain a set of TypedElements. Packages, TypedElements and Types are NamedElements, thus they have a name. Type has the two subclasses Class and Datatype. A Class can optionally be abstract and can be specialized. Classes contain properties which are represented by the class Property. A Property is a MultiplicityElement and TypedElement. This means, for each Property a lower and upper cardinality is defined and each Property has at least one Type. DataTypes can be either PrimitiveTypes or Enumerations, where Enumerations contain a set of literals (not depicted in Figure 2.

Another metametamodel is provided by the Ecore metamodelling language, which is used in the Eclipse Modelling Framework [8]. It is an implementation of EMOF and will be considered in the rest of this paper. Ecore provides four

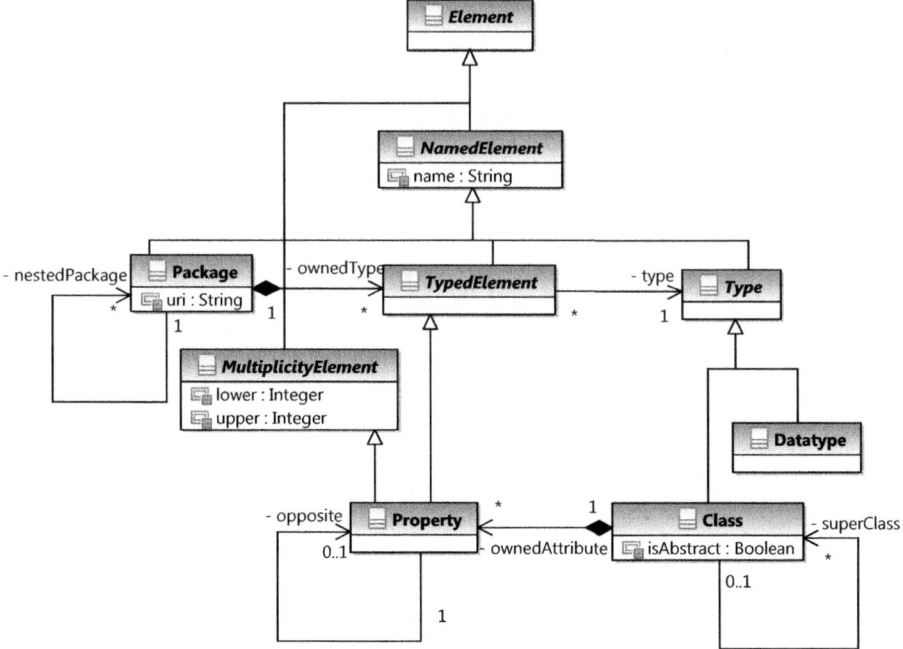

Fig. 2. Essential Meta Object Facility (EMOF)

basic constructs: (1) EClass - used for representing a modeled class. It has a name, zero or more attributes, and zero or more references. (2) EAttribute - used for representing a modeled attribute. Attributes have a name and a type. (3) EReference - used for representing an association between classes. (4) EDataType - used for representing attribute types.

As already mentioned in the introduction, models, metamodels and metametamodels are arranged in a hierarchy of 4 layers. Figure 3 depicts such a hierarchy. Here the Ecore metametamodel is chosen to define a metamodel for a process language, which is built by the language designer. He uses the metametamodel by creating instances of the concepts it provides. The language user takes into account the metamodel and creates instances which build a concrete process model. A process model itself, for example, can describe the behavior of a system running in the real world.

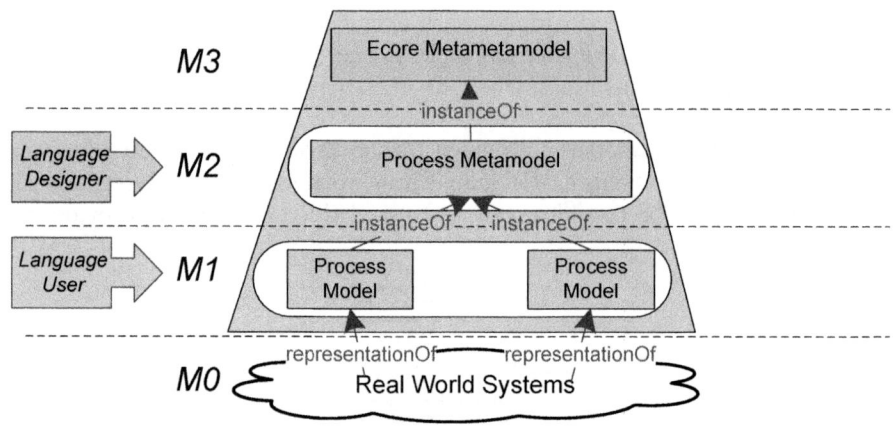

Fig. 3. A metamodel hierarchy

Figure 5 depicts a sample process model in concrete syntax which is instance of the metamodel depicted in Figure 4. Here we used a textual syntax to define classes like ActivityNode or ActivityEdge and references like incoming or outgoing to define links between instances of corresponding classes in the metamodel. In particular we considered the KM3 (Kernel MetaMetaModel) metamodelling language [9] to design the metamodel in figure 4. KM3 is an implementation of EMOF and provides a textual concrete syntax for coding M2 metamodels.

Action nodes (e.g., Receive Order, Fill Order) are used to model concrete actions within an activity. Object nodes (e.g. Invoice) can be used in a variety of ways, depending on where values or objects are flowing from and to.

Control nodes (e.g. the initial node before Receive Order, the decision node after Receive Order, and the fork node and join node around Ship Order, merge node before Close Order, and activity final after Close Order) are used to coordinate the flows between other nodes.

Process models in our example can contain two types of edges, where edges have exactly one source and one target node. One edge is used for object flows and another edge for control flows. An object flow edge models the flow of values to or from object nodes. A control flow is an edge that starts an action or control node after the previous one is finished.

```
1   abstract class ActivityNode {
      reference incoming [0−∗] : ActivityEdge oppositeOf target;
      reference outgoing [0−∗] : ActivityEdge oppositeOf source;
    }
    class ObjectNode extends ActivityNode {  }
6   class Action extends ActivityNode {
      attribute name : String;
    }

    abstract class ControlNode extends ActivityNode {  }
11  class Initial extends ControlNode {  }
    class Final extends ControlNode {  }
    class Fork extends ControlNode {  }
    class Join extends ControlNode {  }
    class Merge extends ControlNode {  }
16  class Decision extends ControlNode {  }

    abstract class ActivityEdge {
      reference source [1−1] : ActivityNode;
      reference target [1−1] : ActivityNode;
21  }
    class ObjectFlow extends ActivityEdge {  }
    class ControlFlow extends ActivityEdge {  }
```

Fig. 4. Process metamodel at M2 layer

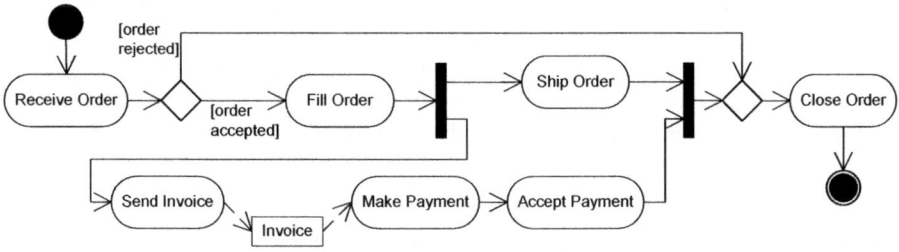

Fig. 5. Process model at M1 layer

2.2 Model Transformations

In the following, we present the idea of model transformations. Figure 6 gives an overview of all relevant artifacts that are involved in defining and executing a model transformation.

A model transformation automatically generates a target model from a source model, according to a transformation definition. Here the source model and the target model conform to the metamodel of the corresponding source and

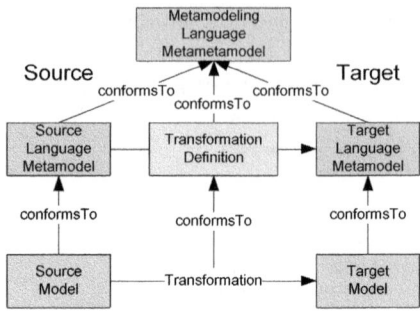

Fig. 6. Model transformations

target language. Further both metamodels conform to a metametamodel of the metamodelling language.

A transformation definition is a set of transformation rules that together describe how a model conforming to the source metamodel can be transformed into a model conforming to the target metamodel. A transformation rule is a description of how one or more constructs in the source metamodel can be transformed into one or more constructs in the target metamodel.

Some prominent model transformation languages are ATL [10], QVT [11], MOLA [12] and FUJABA [13].

2.3 Challenges

In the following, we want to list two MDE challenges which partially come from [14] and are exemplified in the following. Currently, a language designer takes into account a usual metamodelling language which only allows for describing structural aspects of models by concepts like classes or references. Language users consider these languages to create models and require services (e.g. guidance, debugging) for more productive modelling. To provide services to language users, a language designer must specify modelling languages in a more formal way by describing well-formedness constraints in metamodels and (partially) formal semantics of modelling languages.

Metamodel well-formedness constraints and checks. Model correctness is often analyzed implicitly in procedural checks of the modelling tools. To make well-formedness constraints more explicit, they should be defined precisely in the metamodel specification.

Although the Ecore language allows for defining cardinality restrictions for references it is impossible to define constraints covering all flows of a given process model. For example, a language designer wants to ensure, that every flow in an M1 process model goes from the initial node to some final node. To achieve such constraints, language designers have to define formal syntactic constraints of the modelling language which models have to fulfill. Such constraints should be declared directly within the metamodel of the language. Based on metamodel constraints, language designers may want to provide services to the language user.

Language semantics. The semantics of modelling languages is often not defined explicitly but hidden in modelling tools. To fix a specific formal semantics for languages, it should be defined precisely either in the metamodel specification or by transformations which transform software models into logic representations.

This section provided some foundations for software modelling. In the next section we give some foundations on ontology languages and technologies. Later from Section 4 to Section 6 we will show how to use ontology technologies to tackle the challenges mentioned above.

3   Ontology Languages and Technologies

In this section, we start with the metamodel of the ontology language OWL 2. Furthermore, we are dealing with ontology-based reasoning services which are later used in designing and using modelling languages. We start with a list of standard reasoning services (Section 3.2). Additionally, we consider in this section querying services which are based on SPARQL (Section 3.3), explanation services to explain inferences in ontologies (Section 3.4), repairing services which give advices of how to correct ontologies (Section 3.5), and linking services which show how to combine different ontologies (Section 3.6).

3.1   Ontology Language OWL

Improvements on the OWL language led the W3C OWL Working Group to publish working drafts of a new version of OWL: OWL 2.

OWL 2 is fully compatible with OWL-DL and extends the latter with limited complex role inclusion axioms, reflexivity and irreflexivity, role disjointness and qualified cardinality restrictions. Moreover, OWL 2 is axiom-based. The three important kinds of axioms are class axioms, object property axioms and data property axioms. Class axioms like SubClassOf- or EquivalentClasses-axioms are combined with class expressions and describe a subclass relation between two classes and a set of classes with equivalent concepts, respectively. Two of the object property axioms are the ObjectPropertyDomain- and ObjectPropertyRange-axioms which restrict the object property only to be connected with given class expressions defined in the domain and range. The data property axioms for the data property domain are analogous to the one of the object property. The axioms for data property range define the data range a given data property contains. Classes and properties themselves are entities and are declared by a corresponding Declaration-axiom.

The complete abstract syntax is presented in [3]. Its semantics is presented in [15]

3.2   Standard Ontology Reasoning Services

In the following paragraphs, we consider standard reasoning services that are provided by reasoners (e.g., Pellet [16]).

Consistency Checking. The reasoning service consistency checking checks if a given ontology $\mathcal{O}$ is consistent, i.e. if there exists a model (a model-theoretic instance) for $\mathcal{O}$. If ontology $\mathcal{O}$ is consistent, then return *true*, otherwise *false*.

Satisfiability Checking. The satisfiability checking service finds all unsatisfiable concepts in a given ontology $\mathcal{O}$. A concept in an ontology $\mathcal{O}$ is unsatisfiable if it represents an empty set of all models in the ontology, i.e. it cannot be instantiated. If the ontology $\mathcal{O}$ has no unsatisfiable concept the service returns the empty set.

Classification. The classification service returns for a given ontology $\mathcal{O}$ and an individual i a set of concepts which contain/describe the individual. The individual conforms to all concepts in the result of the classification service.

Subsumption. The subsumption checking service checks whether the interpretation of A, the set of individuals described by A, is a subset of the interpretation of B for a given ontology $\mathcal{O}$. If the interpretation of A is a subset of the interpretation of B, then it returns *true*. Otherwise it returns *false*.

3.3   Ontology Querying Service

SPARQL is the W3C standard query language for RDF graphs, which is a triple-based language [17]. Writing SPARQL queries for OWL can be time-consuming for those who work with OWL ontologies, since OWL is not triple-based and requires reification of axioms when using a triple-based language. Therefore, we proposed SPARQLAS, a language that allows for specifying expressions that rely on inferences over OWL class descriptions [18]. It is a seamless modification of the SPARQL syntax for querying OWL ontologies. SPARQLAS enables using variables (prefixed by ?) wherever an entity (Class, Datatype, ObjectProperty, DataProperty, NamedIndividual) or a literal is allowed.

SPARQLAS queries are translated into SPARQL queries and can be executed by any SPARQL engine that supports graph pattern matching for the OWL 2 entailment regime [19]. SPARQLAS queries operate on both the schema level (M2 metamodel layer) and on the instance level (M1 model layer). For example, Listing 1.1 shows a SPARQLAS query about activities that are followed by a decision node. In this example, we ask about the individuals ?x whose type is an anonymous class where the property target has as value some control flow with some property target having some Decision.

Listing 1.1. Use Cases that includes some other use case

```
1  Namespace: uml = <http://www.example.org/BPM#>
   Select ?x
   Where:
       ?x type (ActivityNode and outgoing some (ControlFlow and
           target some Decision))
```

3.4 Explanation Service

In traditional ontology development environments (e.g., Protégé [20]), users are typically able to model ontologies and use reasoners (e.g., Pellet [16]) to compute unsatisfiable classes, subsumption hierarchies and types for individuals.

In the following, we introduce the terminology for justifications and present some ideas on how to compute justifications.

Computing Justifications. Explanation services for MDE that are based on ontology technologies are mainly based on computing so-called justifications.

However, since in the context of MDE ontology technologies are used in software modelling it has become evident that there is a significant demand for software modelling environments which provide more sophisticated explanation services. In particular, the generation of explanations, or justifications, for inferences computed by a reasoner is now recognized as highly desirable functionality for both ontology development and software modelling. A language user or designer developing (meta-) model recognizes an entailment and wants to get an explanation for the entailment in order to get the reason why the entailment holds (understanding entailments). If the entailment leads to some inconsistency or unsatisfiable classes, the user wants to get some debugging relevant facts and the information how to repair the ontology.

In the following, we present some ideas on how to compute justifications.

Terminology. In the following, the terminology that is related to the field of justification and debugging is depicted.

A class is unsatisfiable (with regard to one ontology) if it cannot possibly have any instances in any model of the ontology. In description logics notation, $C \sqsubseteq \bot$ means that C is not satisfiable. If an ontology $\mathcal{O}$ contains at least one unsatisfiable class it is called *incoherent*. An ontology $\mathcal{O}$ is inconsistent if and only if it does not have any model. In description logics, this is the case if an ontology $\mathcal{O}$ entails $\top \sqsubseteq \bot$. $\mathcal{O} \models \eta$ holds if all models of $\mathcal{O}$ also satisfy η. Justifications are explanations of entailments in ontologies. Let $\mathcal{O}$ be an ontology with entailment $\mathcal{O} \models \eta$. Then $\mathcal{J}$ is a justification for η if $\mathcal{J} \subseteq \mathcal{O}$ with $\mathcal{J} \models \eta$ and for any $\mathcal{J}' \subset \mathcal{J}$ the following holds: $\mathcal{J}' \not\models \eta$.

In the following, we present the ideas of a simple black box method for computing a single justification and two further methods for computing more than one possible justification. For details about the algorithms we refer to literature, e.g., [21,22].

Simple Black Box Method. The approach of a simple black-box technique to compute justifications was presented in [22]. Given a concept C which is unsatisfiable with regard to an ontology $\mathcal{O}$. In a first step of the computation, axioms of $\mathcal{O}$ are added to a newly created ontology $\mathcal{O}'$ until C gets unsatisfiable with regard to $\mathcal{O}'$. In a second step extraneous axioms in $\mathcal{O}'$ will be deleted to get a single minimal justification. The deletion of axioms stops when concept C gets satisfiable.

Glass Box Method. In [21] a glass box technique for computing one single justification is presented. The technique is used for presenting the root cause of a contradiction and to determine the minimal set of axioms in the ontology which lead to a semantic clash. In glass box techniques, the internals of a description logics tableaux reasoner are modified to extract and reveal the cause for inconsistency of a concept definition. An advantage of such approaches is that by tightly integrating the debugging with the reasoning procedure, precise results can be obtained. On the other hand, the reasoner needs to maintain extra data structures to track the source and its dependencies and this introduces additional memory and computation consumption. We refer to [21] where the complete algorithms and methodologies are explained in detail.

Computing all justifications. If an initial justification is given (for example, computed by some black box technique), other techniques are used to compute the remaining ones. In [22] a variation of the classical Hitting Set Tree (HST) algorithm [23] is presented. This technique is also reasoner independent (blackbox). The idea is that given an algorithm to find a single justification for concept unsatisfiability (like the above black-box method), to find in a first step one justification and set it as the root node of the so called Hitting Set Tree (HST). In the next steps, each of the axioms in the justification is removed individually, thereby creating new branches of the HST, and find new justifications along these branches on the fly in the modified ontology. This process needs to be exhaustively done in order to compute all justifications. The algorithm repeats this process until the concept turns satisfiable.

Example. Listing 1.2 depicts an ontology represented in the textual functional-style syntax. The ontology consists of concepts of the domain of physical devices. The general physical structure of a Device consists of a Configuration which has a number of Slots into which Cards can be plugged in. (In Section 5 we will consider the domain of physical devices in the context of MDE where domain-specific languages are used to model devices and its possible configurations.)

The most important class Device has as superclass an anonymous OWL concept, which defines that every device is connected via the property hasConfig with Configuration7603. Furthermore, class Device is equivalent with an anonymous concept which requires, that each device is connected via property hasConfig with some Configuration7604.

The property hasConfig connects Device with the intersection of the classes Configuration7603 and Configuration7604.

Further classes in the metamodel are Cisco, a subclass of Device and class Configuration, which has two subclasses, namely Configuration7603 and Configuration7604 which are pairwise disjoint.

It is obvious that the ontology contains two unsatisfiable classes, namely Device and Cisco. The objectives for the explanation service are to provide a set of justifications, consisting of axioms. To get an explanation, why the class Device is not satisfiable, the metamodel is transformed into a description logics TBox (as explained in Section 5).

Listing 1.2. Example of Metamodel enriched by ontology expressions

```
1 EquivalentClasses(Device ObjectSomeValuesFrom(hasConfig Configuration7603))
  SubClassOf(Device ObjectSomeValuesFrom(hasConfig Configuration7604))

  SubClassOf(Cisco Device)
5
  SubClassOf(Configuration7604 Configuration)
  DisjointClasses(Configuration7604 Configuration7603)

  SubClassOf(Configuration owl:Thing)
10
  SubClassOf(Configuration7603 Configuration)
  DisjointClasses(Configuration7603 Configuration7604)

  ObjectPropertyDomain(hasConfig Device)
15 ObjectPropertyRange(hasConfig ObjectIntersectionOf(Configuration7604 Configuration7603))
```

In Listing 1.3 we present two justifications for class Device and class Cisco, respectively, which are for simpler reading and understanding rendered using the Manchester Syntax style [24]. The output was generated by the TwoUse Toolkit(cf. Section 7).

Listing 1.3. Justifications for metamodel unsatisfiability

```
1 Unsatisfiability of Device:
  Explanations (2):
  1) hasConfig range Configuration7603
                  and Configuration7604
5     Configuration7603 disjointWith Configuration7604
      Device subClassOf hasConfig some Configuration7604

  2) hasConfig range Configuration7603
                  and Configuration7604
10    Configuration7603 disjointWith Configuration7604
      Device equivalentTo hasConfig some Configuration7603
  _ _ _ _ _ _ _ _ _ _ _ _ _ _ _ _ _ _ _ _ _ _ _ _ _ _ _ _
  Unsatisfiability of Cisco:
  Explanations (2):
15 1) hasConfig range Configuration7603
                  and Configuration7604
      Configuration7603 disjointWith Configuration7604
      Cisco subClassOf Device
      Device subClassOf hasConfig some Configuration7604
20
  2) hasConfig range Configuration7603
                  and Configuration7604
      Configuration7603 disjointWith Configuration7604
      Cisco subClassOf Device
25    Device equivalentTo hasConfig some Configuration7603
```

3.5 Ontology Repair Service

In the above section, we have highlighted some ontology explanation services that can be used to highlight the core erroneous axioms and concepts in a defect ontology. The ontology results from the language metamodel and model, which are transformed into the ontology. The next step is to resolve the errors by processing the justifications to give at least advices how to repair the ontology, respectively metamodel and model.

In the following, we discuss an idea for computing advices for repairing/changing models. In general, approaches for ontology repairing are presented by Kalyanpur et al., for example, in [25,26].

Although there might be (semi-)automatic strategies for repairing ontologies and models, each step underlies the knowledge and proof of domain experts.

Based on a set of justifications different axiom rating strategies can be adopted. A simple one is to compute the frequency of an axiom. Here the number of times the axiom appears in each justification of the various unsatisfiable concepts in an ontology is counted. If an axiom appears in all justifications for n different unsatisfiable concepts removing the axiom from the ontology ensures that n concepts become satisfiable. Thus, the higher the frequency, the lower (better) the rank assigned to the axiom.

With regard to our example with the two unsatisfiable classes Device and Cisco, the axioms Configuration7603 disjointWith Configuration7604 and hasConfig range Configuration7603 and Configuration7604 have the highest frequencies. Thus, removing one of the axioms would solve the unsatisfiability problem of class Device and Cisco.

3.6   Ontology Linking Service

In a model-driven paradigm, resources that are expressed using different modelling languages must be reconciled before being used. To apply (semi-) automatically linking between models and metamodels, both are transformed into an ontology, where models are transformed into the description logics ABox and metamodels to the description logics TBox. In this section, we only illustrate some of the multiple ontology matching techniques. For a deeper understanding on this topic, please refer to [27].

Ontology matching is the discipline responsible for studying techniques for reconciling multiple resources on the web. It consists of two steps: (1) match and determine alignments and (2) the generation of a processor for merging and transforming the ontologies. Matching identifies the correspondences. A correspondence for two ontologies A and B is a quintuple including an id, an entity of ontology A, an entity of ontology B, a relation (equivalence, more general, disjointness) and a confidence measure. A set of correspondences is an alignment. Correspondences can be done at the schema-level (metamodel) and at the instance-level (model).

Matchings can be based on different criteria: name of entities, structure (relations between entities, cardinality), background knowledge like existing ontologies or WordNet. Furthermore, matchings are established according to the different structures that are compared.

Internal structure comparison: this includes property, key, datatype, domain and multiplicities comparison.

Relational structure comparison: the taxonomic structure between the ontologies is compared.

Extensional techniques: extensional information is used in this method, e.g., formal concept analysis.

Automatic matching techniques can be seen as support but should be assisted by domain experts, because of false positive matches.

In the next Section 4 we will show how to adapt ontology languages and ontology technologies on software modelling. Here we present two generic approaches which will be examplified in Section 5 and Section 6.

4 Bridging Software Languages and Ontology Technologies

In the following, we present two general approaches of bridging software models and ontologies. The two approaches mainly differ in the layer of the model hierarchy where they are defined and the layer where the bridge is used and applied on software models.

4.1 Language Bridge

Figure 7 depicts the general architecture of a language bridge, combining software languages and ontology technologies. The bridge itself is defined at the M3 layer, where a metametamodel like Ecore is considered and bridged with the OWL metamodel. Here we differ between two kinds of bridges: *M3 Integration Bridge* and *M3 Transformation Bridge*.

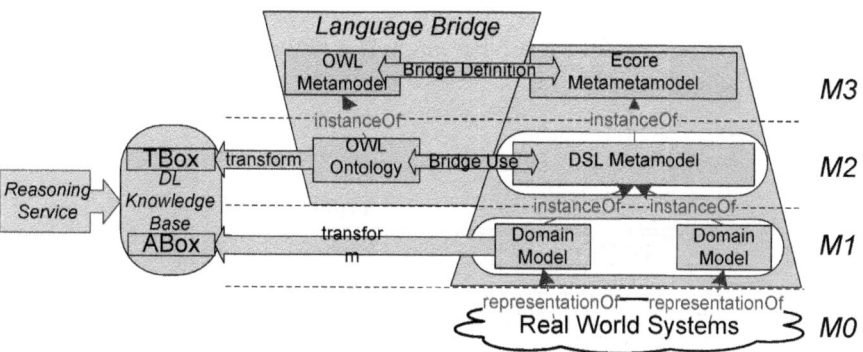

Fig. 7. Language bridge

M3 Integration Bridge The design of an M3 integration bridge consists mainly of identifying concepts in the Ecore metametamodel and the OWL metamodel which are combined.

Here existing metamodel integration approaches (e.g. presented in [28] and [29]) to combine the different metamodels are used. The result is a new metamodelling language, which allows for designing language metamodels at the M2 layer with integrated constraints.

An integrated metamodelling language provides all classes of the Ecore metametamodel and OWL metamodel. It merges, for example, OWL Class with Ecore EClass, OWL ObjectProperty with Ecore References or OWL DataProperty with Ecore Attribute. Thus, a strong connection between the two languages is built. Since a language designer creates a class, he is in the scope of both OWL class and ECore class. Hence a language designer can use the designed class within OWL class axioms and simultaneously use features of the Ecore metamodelling language, like the definition of simple references between two classes.

The integration bridge itself is used at the M2 layer by a language designer. He is now able to define language metamodels with integrated OWL annotations to restrict the use of concepts he modeled and to extend the expressiveness of the language.

To provide reasoning services to language users and language designers, the integrated metamodel is transformed into a Description Logics TBox. The models created by the language users are transformed into a corresponding Description Logics ABox. Based on the knowledge base consisting of a TBox and ABox we can provide standard reasoning services and provide specific modelling to both language user and designer.

M3 Transformation Bridge. The M3 Transformation Bridge allows language designers and language users to achieve representations of software languages (Metamodel/Model) in OWL. It provides the transformation of software language constructs like classes and properties into corresponding OWL constructs.

As one might notice, Ecore and OWL have a lot of similar constructs like classes, attributes and references. To extend the expressiveness of Ecore with OWL constructs, we need to establish mappings between the Ecore constructs onto OWL constructs. Table 4.1 presents a complete list of similar constructs.

Table 1. Ecore and OWL: comparable constructs

Ecore	OWL
package	ontology
class	class
instance and literals	individual and literals
reference, attribute	object property, data property
data types	data types
enumeration	enumeration
multiplicity	cardinality

Based on these mapping, we develop a generic transformation script to transform any Ecore Metamodel/Model into OWL TBox/ABox – *OWLizer*. Figure 8 depicts the conceptual schema of transforming Ecore into OWL.

Figure 8 shows three modelling levels according to the OMGs metamodel architecture: the metametamodel level (M3), the metamodel level (M2) and the model level (M1). Vertical arrows denote instantiation whereas the horizontal arrows are transformations, and boxes represent packages.

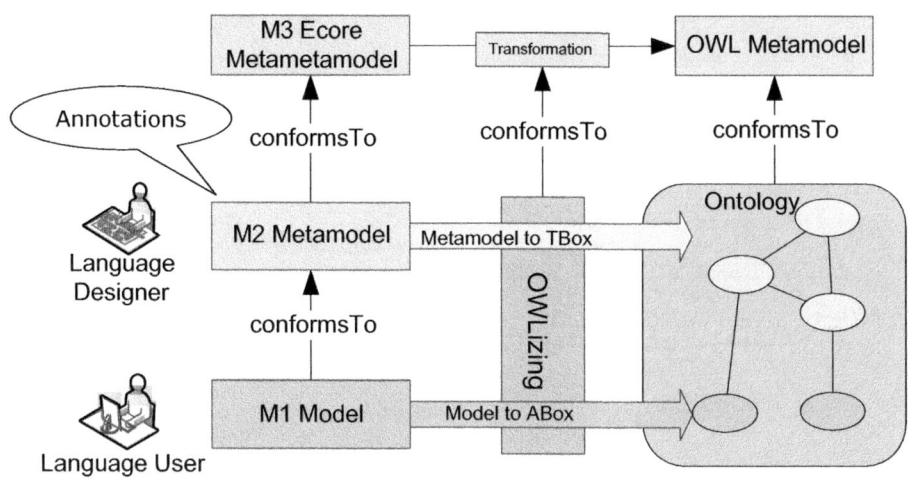

Fig. 8. OWLizer

A model transformation takes the UML metamodel and the annotations as input and generates an OWL ontology where the concepts, enumerations, properties and data types (TBox) correspond to classes, enumerations, attributes/references and data types in the UML metamodel. Another transformation takes the UML model created by the UML user and generates individuals in the same OWL ontology. The whole process is completely transparent for UML users.

As one may notice, this is a generic approach to be used with any Ecore-based language. For example, one might want to transform the UML Metamodel/Models as well as all the Java grammar/code into OWL (classes/individuals). This approach can be seen as a linked data driven software development environment [30].

4.2 Model Bridge

Model bridges connect software models and ontologies on the modelling layer M1. They are defined in the metamodelling layer M2 between different metamodels. Figure 9 visualises a model bridge. The bridge is defined between a process metamodel on the software modelling side and an OWL metamodel in the OWL modelling hierarchy. The process metamodel is an instance of an Ecore (EMOF) metametamodel.

A model bridge is defined as follows: (1) Constructs in the software modelling and in the ontology space are identified. These constructs, or language constructs, are used to define the corresponding models in the modelling layer M1. (2) Based on the identification of the constructs, the relationship between the constructs are analyzed and specified, i.e. the relationship of an Activity in a process metamodel like the BPMN metamodel to an OWL class. We distinguish between a *transformation* and *integration bridge*.

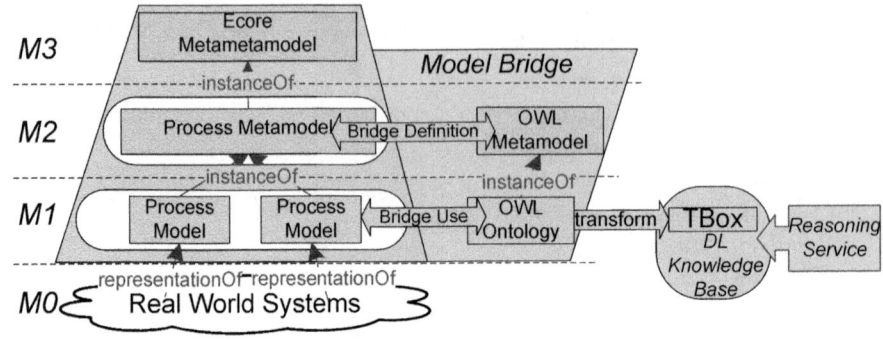

Fig. 9. Model bridge

M2 Integration Bridge *Integration bridges* merge information of the models from the software modelling and from the ontology space. This allows the building of integrated models (on modelling layer M1) using constructs of both modelling languages in a combined way, e.g. to integrate UML class diagrams and OWL.

As mentioned in Section 4.1, UML class-based modelling and OWL comprise some constituents that are similar in many respects like classes, associations, properties, packages, types, generalization and instances [31]. Since both approaches provide complementary benefits, contemporary software development should make use of both. The benefits of an integration are twofold. Firstly, it provides software developers with more modelling power. Secondly, it enables semantic software developers to use object-oriented concepts like inheritance, operation and polymorphism together with ontologies in a platform independent way.

Such an integration is not only intriguing because of the heterogeneity of the two modelling approaches, but it is now a strict requirement to allow for the development of software with many thousands of ontology classes and multiple dozens of complex software modules in the realms of medical informatics [32], multimedia [33] or engineering applications [34].

TwoUse (Transforming and Weaving Ontologies and UML in Software Engineering) addresses these types of systems [35]. It is an approach combining UML class-based models with OWL ontologies to leverage the unique and potentially complementary strengths of the two. TwoUse consists of an integration of the MOF-based metamodels for UML and OWL, the specification of dynamic behavior referring to OWL reasoning and the definition of a joint profile for denoting hybrid models as well as other concrete syntaxes.

Figure 10 presents a model-driven view of the TwoUse approach. TwoUse uses UML profiled class diagrams as concrete syntax for designing combined models. The UML class diagrams profiled for TwoUse are input for model transformations that generate TwoUse models conforming to the TwoUse metamodel. The TwoUse metamodel provides the abstract syntax for the TwoUse approach, since

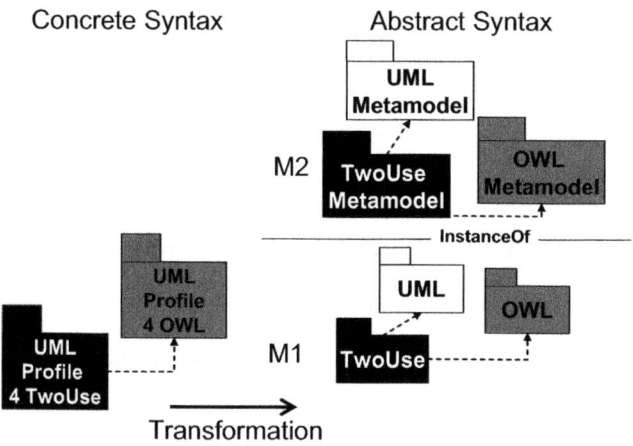

Fig. 10. Example of M2 Integration Bridge

we have explored different concrete syntaxes. Further model transformations take TwoUse models and generate the OWL ontology and Java code.

TwoUse allows developers to raise the level of abstraction of business rules previously embedded in code. It enables UML modelling with semantic expressiveness of OWL DL. TwoUse achieves improvements on the maintainability, reusability and extensibility for ontology based system development.

M2 Transformation Bridge. A *transformation bridge* describes a (physical) transformation between models in layer M1. The models are kept separately in both modelling spaces. The information is moved from one model to the model in the other modelling space according to the transformation bridge. With respect to the example depicted in Figure 9, a process model like a UML Activity Diagram is transformed to an OWL ontology. The transformation rules or patterns are defined by the bridge. Thus, having a process model as an ontology we can provide services for reasoning on the semantics of process models.

5   Ontology Reasoning for Structural Modelling

In this section, we show how bridge (domain-specific) modelling languages with ontology languages. In particular, we show how the abstract syntax of modelling languages can be restricted by the use of integrated ontology languages. The challenge is to formally define language metamodels with integrated ontology-based axioms and expressions that allow for reasoning on the metamodel itself and all conforming models.

In recent works (e.g. [36,37]) we have exemplified, that ontology reasoning for structural modeling is mainly based on combining a metametamodel (e.g. Ecore) with the metamodel of an ontology language. Having a metametamodel bridge with the OWL metamodel a language designer is able, for example, to define formal well-formedness constraints based on OWL to restrict the use of concepts provided by the metamodel. The metamodel and conforming models are transformed to an ontology representation for reasoning on the structure of models that is prescribed by the language metamodel (and additional well-formedness constraints).

5.1   Challenges and Tasks in Structural Modelling

Comarch[1], one of the industrial partners in the *MOST project*[2] has provided the running example used in this section. It is an excerpt of a use case where telecommunication providers want to model configurations for telecommunication systems.

Let us elaborate the following example: The general physical structure of a Device consists of a Bay which has a number of Shelfs. A Shelf contains Slots into which Cards can be plugged. Logically, a Shelf with its possible Slots and Cards is stored as a Configuration.

Figure 11 depicts the development of a domain model for physical devices by a language user. Firstly (*step 1*), the language user starts with an instance of the general concept Device. A device requires at least one configuration. Thus the language user plugs in a Configuration element into the device.

In *step 2*, the language user adds exactly three slots to the device model. At this point, the language user wants to verify whether the configuration satisfies the domain restrictions, which is done, for example, by invoking a query against the current physical device model.

After adding three slots to the model of the physical device, the language user plugs in some cards to complete the end product (*step 3*). Knowing which cards and interfaces should be provided by the device, he may insert an SPA Interface Card for 1-Gbps broadband connections, a Supervisor Engine 720 card for different IP and security features and a controller for swapping cards at runtime (Hot Swap Controller). At this point, the language user wants to use reasoning services.

The domain-specific language (DSL) defines the knowledge about which special types of cards are provided by a Configuration. Having the information that its instance is connected with three slots, the replacement of the Configuration type by the more specific type Configuration7603 is recommended to the language user as result of activating the reasoning service. Moreover, the language user is informed how this suggestion takes place and about restrictions related to such a configuration.

Since it has been inferred that the device has the Configuration7603, in *step 4*, the available reasoning service for the Device element infers that the device is one of type Cisco7603. The necessary and sufficient condition to be a Cisco7603 are checked by reasoning services.

---

[1] http://www.comarch.com/

[2] http://www.most-project.eu/

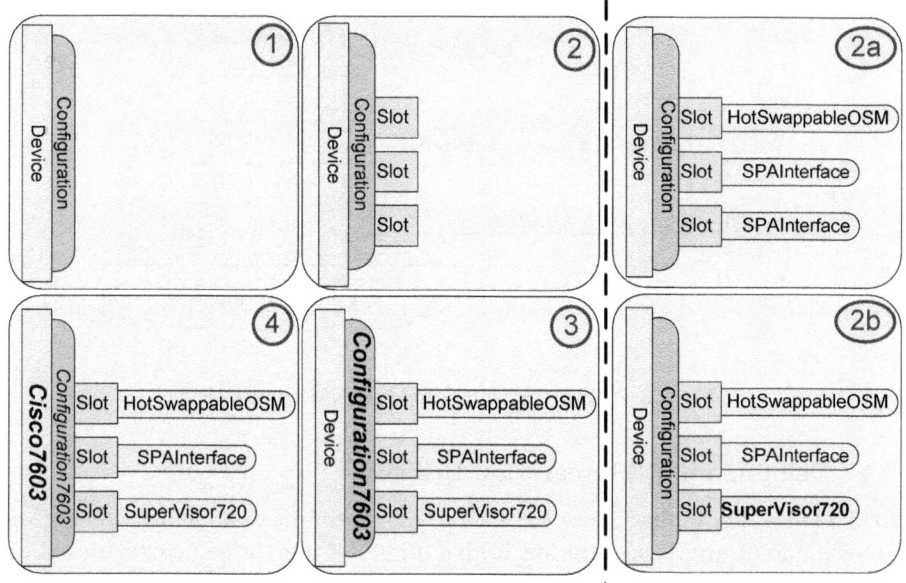

Fig. 11. Modelling a physical device in four steps (M1 layer)

Steps *2a and 2b* show a second path in the scenario of modelling a physical device where debugging comes into play. After creating elements for a device, a configuration and slots, the language user plugs into one slot a HotSwappableOSM card and into the remaining slots two SPAInterface cards (*step 2a*). Here an inconsistency occurs. The language user needs an explanation why there is a problem with this configuration. Explanation services give the information that each configuration must have a slot in which a SuperVisor720 card is plugged in. Having a correct configuration, the DSL user can continue with *steps 3 and 4* as described above.

5.2   Bridges for Structural Modelling Languages

As a bridge for structural modelling we propose an M3 integration bridge, where the bridge is defined at the M3 layer by an integration of a metametamodel and ontology language (cf. Figure 7 and 12).

In particular, we consider an M3 integration bridge where the metamodeling language KM3 [9] is combined with the OWL2 metamodel. An overview of bridge definition and use is depicted in Figure 12.

The bridge itself is used at the M2 layer. Here the language designer is able to define constraints and expressions within his metamodel to formally define the abstract syntax of a new modelling language. The modelling language itself is used at the M1 layer. The constraints and expressions restrict the creation of domain models which are instances of the metamodel.

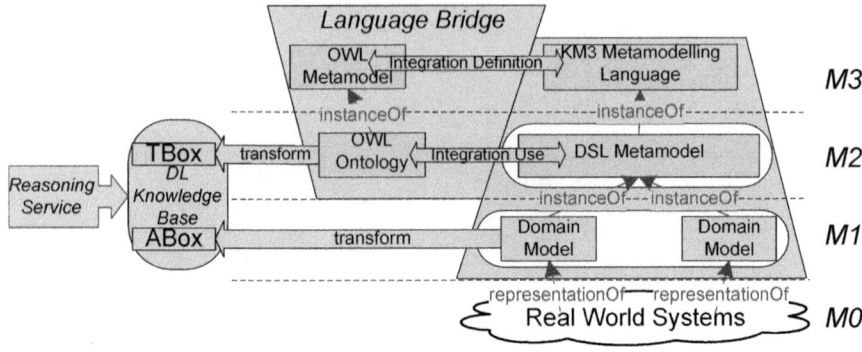

Fig. 12. M3 Integration Bridge on KM3 and OWL

5.3   Defining an M3 Integration Bridge

In the following, we introduce the idea of integrated metamodeling which gives an example of an M3 integration bridge. First we present some excerpts of an M3 metametamodel, which defines the abstract syntax for metamodels with integrated ontology annotations.

Here, we integrate the existing KM3 [9] metametamodel, a simplified subset of EMOF, with the OWL 2 metamodel at the M3 layer. Thus, we can provide a new metamodeling language which allows for integrated modeling of both KM3-based metamodel and OWL ontologies.

Furthermore, we present some parts of the concrete syntax that are used by the language designer to implement metamodels. Like in Section 2.1, we use a textual syntax that is easy to implement, as an extension of the concrete syntax of KM3, and might provide some productivity in coding metamodels. Thus we give the idea of how to define such a syntax using EBNF grammar rules.

Abstract Syntax. The integrated M3 metametamodel enables language designers to describe M2 classes together with OWL annotations. The KM3 metametamodel allows for specifying behavioral and structural features of classes. The OWL metamodel provides for the use of OWL primitives and corresponding sound and complete reasoning over these primitives.

Our integration approach considers classes from the OWL2 metamodel and a further (meta-) metamodel and combines corresponding classes by creating a so-called TwoUSE-class which is a specialization of both classes and thus inherits the properties of both.

Figure 13 depicts an excerpt of the integrated metametamodel. Here, the central concepts are the TUClass, the TUAttribute, and the TUReference.

The integration between the KM3 and the OWL 2 metamodels is done by applying the class adapter design pattern [38] into KM3 metaclasses that are similar to OWL 2 metaclasses (for similarities between OWL and class-based modelling languages see [39]; for integration methods of (meta-)modelling language and ontology languages see [28,40]).

TUClass is a specialization of KM3 Class and OWLClass and inherits their behavior. Using the TwoUses classes TUAttribute and TUReference we connect OWL DataProperty with KM3 Attribute and OWL ObjectProperty with KM3 Reference respectively. Using the TwoUse classes for references and attributes of KM3 we ensure that all instances have the behavior of an OWL Object Property or OWL Data Property respectively.

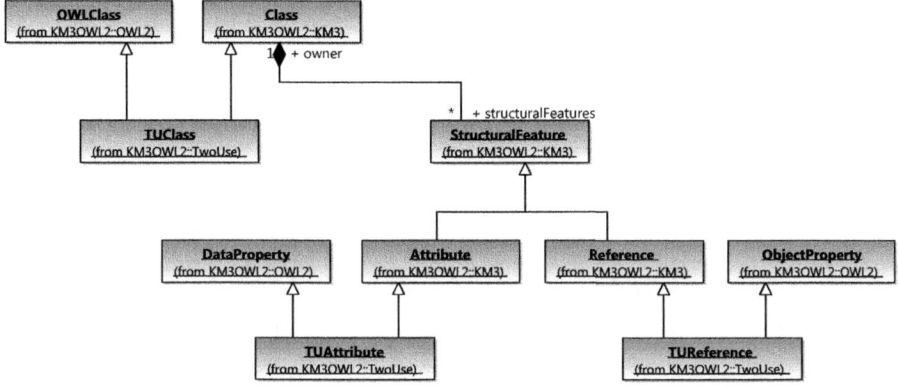

Fig. 13. Excerpt of KM3+OWL 2 metametamodel (M3 layer)

Concrete Syntax. In the following, we present some parts of a concrete syntax which extends the one from KM3 [9]. In this section, we only want to give the idea of how to extend the KM3 syntax by OWL class axioms and object property axioms. Because the KM3 concrete syntax is based on a grammar, in the following we define rules in EBNF and give an example of using them to code metamodels.

OWL Class Axioms. The first line of Listing 1.4 shows rules for the definition of a class. Each class is optionally abstract, is defined by the keyword class, has a name and a list of super types. Between the curly brackets a list of class features are defined such as attributes and references. In our special case the grammar rule is extended by the new non-terminal classAxioms which optionally produces a list of OWL 2 class axioms (see Listing 1.4, line 2). The non-terminal ClassAxiom is used and is defined by an OWL 2 textual concrete syntax.

Listing 1.4. Adopting OWL Class Axioms

```
1 class = ["abstract"] "class" name [supertypes] [classAxioms] "{" features "}";
  classAxioms = ClassAxiom { "," ClassAxiom};
```

Listing 1.5 presents an example of using OWL class axioms in combination with the declaration of classes in KM3. In line 1, the class Configuration is defined having a reference called hasSlot to a further class Slot. Line 4 depicts the head of the class Configuration7603 which has the super type Configuration and thus inherits the feature hasSlot. After declaring the super types the definition of an OWL class axiom follows. It defines that the class Configuration7603 is equivalent with the anonymous OWL class restrictionOn hasSlot with exactly 3 Slot, the instances which are connected with exactly 3 further instances of type Slot using the hasSlot reference.

Listing 1.5. Example of using OWL Class Axioms (M2 layer)

```
1 class Configuration extends restrictionOn hasSlot with min 1 Slot {
    reference hasSlot [1−*]: Slot;
  }
  class Configuration7603 extends Configuration equivalentTo restrictionOn hasSlot with exactly 3 Slot {
5 }
```

Object Property Axioms. Listing 1.6 presents the grammar rules that adopt OWL object property axioms on references. A general reference feature in KM3 is declared by the keyword reference, a name, a multiplicity, the optional indication if the reference acts as a container, and a type where the reference points to. In addition to this declaration, we introduce the new non-terminal object-PropertyAxioms which is defined in line 2 of the Listing and produces a list of ObjectPropertyAxiom. Again this non-terminal is defined in our OWL 2 natural style syntax. Thus a list of OWL object property axioms can be append to the declaration of KM3 references.

Listing 1.6. Adopting OWL Object Property Axioms

```
1 reference = "reference" name multiplicity isContainer ":" typeref "oppositeOf"
  name [objectPropertyAxioms] ";";
  objectPropertyAxioms = ObjectPropertyAxiom { "," ObjectPropertyAxiom };
```

Listing 1.7 depicts an example of using OWL object property axioms together with the reference features of KM3. Here the two classes Slot and its subclass Slot7609-2 are defined. Slot contains a reference called hasCard, Slot7609-2 contains a reference called hasInterfaceCard. hasCard points to elements of type Card, hasInterfaceCard points to elements of type CiscoInterface, a subclass of Card. Using OWL object property axioms we state that hasInterfaceCard is a sub property of hasCard. Thus, if an instance of CiscoInterface is connected via the reference hasInterface with some interface we can infer that this instance is also connected via the reference hasCard with the interface instance. Thus, the restriction Min-Cardinality(1 hasCard) holds.

Listing 1.7. Example of using OWL Object Property Axioms (M2 layer)

```
1 class Slot equivalentTo restrictionOn hasCard with min 1 Card{
    reference hasCard [1−*]: Card;
  }

5 class Slot7609−2 extends Slot{
    reference hasInterfaceCard [1−*]: CiscoInterface subpropertyOf hasCard;
  }
```

5.4   Using an M3 Integration Bridge

In the following, we show how to adopt reasoning services on structural modelling languages that are used for modelling configurations. In particular, we consider languages for modelling configurations of physical devices as introduced in Section 5.1. In the following, we examplify the bridging approach which tackles the design and use of the PDDSL (Physical Device DSL). Further bridging approaches between configuration languages and ontology technologies can be found in [41].

Designing PDDSL. Listing 1.8 depicts an example of an integrated PDDSL language. Here, we define constraints and restrictions within the metamodel definition.

To design the metamodel of PDDSL we used the combined metamodelling language consisting of KM3+OWL. The metamodel represent the abstract syntax of the PDDSL as well as well-formedness constraints for the *M1-layer*. This additional constraints are useful to define the syntactic structure of the domain model at the M1-layer as well as to indicate constraints that apply at the level of the modelling language itself (M2-layer).

Listing 1.8. Example of defining an integrated metamodel for PDDSL

```
 1 class Device {
     reference hasConfiguration [1−∗]: Configuration;
   }

 5 class Cisco7603 extends Device, equivalentWith restrictionOn hasConfiguration
   with min 1 Configuration7603 {
   }

   class Configuration extends IntersectionOf(restrictionOn hasSlot with min 1
10 Slot, restrictionOn hasSlot with some restrictionOn hasCard with some
   SuperVisor720){
     reference hasSlot : Slot;
   }

15 class Configuration7603 extends Configuration, equivalentWith
   IntersectionOf(restrictionOn hasSlot with exactly 3 Slot, restrictionOn hasSlot
   with some restrictionOn hasCard with some UnionOf(HotSwappableOSM,
   SPAinterfaceProcessors) { }

20 class Slot {
     reference hasCard [1−∗]: Card;
   }

   class Card {
25 }

   class SuperVisor720 extends Card {
   }

30 class SPAinterfaceProcessors extends Card {
   }

   class HotSwappableOSM extends Card {
   }
```

Using PDDSL. Having a PDDSL metamodel specified by the language designer, the language user is able to build different domain models. These domain models describe possible configurations of physical network devices. Possible domain models are depicted in Figure 11. During domain modelling, the language user experiences several benefits. In development environments these benefits would be implemented by reasoning services (cf. Section 3) which are automatically invoked. The language user needs no background information on how the reasoning services work and how they are connected with the knowledge base. The reasoning engine returns suggestions and explanations to the DSL user.

To allow for such services, the ABox and TBox of an ontology are extracted from the domain model (*M1 layer*) and the integrated metamodel (*M2 layer*), respectively. In the following, we will consider more precisely the services that are provided to user.

Reasoning Services for Modelling Configurations. In the following, we present some concrete services that are used by language users.

Detecting Inconsistencies in Domain Models. To detect inconsistencies in domain models the PDDSL metamodel is transformed into a description logics TBox. All classes in the PDDSL metamodel are transformed to a concept in a description logics TBox. All references in the metamodel are represented by roles (object properties) in the TBox. The domain model itself is transformed to a description logics TBox to check whether the domain models are consistent we use the *consistency checking reasoning service.* With regard to the example the service returns that the model in Figure 11 (2a), is inconsistent, because the configuration does not contain the mandatory supervisor card (as it is defined in the metamodel in Figure 1.8).

Finding and Explaining Errors in Configuration Instances. Language users that use the PDDSL language (which describes sets of possible configurations) to create domain models (which describe concrete configurations) require debugging of domain models and explanation of errors and inconsistencies. More precisely, a user of PDDSL wants to identify illegal configurations, wants to get explanations why the configuration of a device is inconsistent and wants to get suggestions how to fix the configuration.

To validate configuration models against the PDDSL metamodel, the model is transformed into a description logics ABox. The TBox is built by the PDDSL metamodel which contains all metaconcepts and additional constraints. The TBox and ABox build the description logics knowledge base. Using the consistency checking reasoning service, we can check if the knowledge base is consistent and thus validate the configuration

If we want to detect invalid cards in a configuration, we need some explanations why the model is inconsistent with regard to its metamodel. Explanations are provided by explanation services introduced in Section 3.4.

Figure 11 (2a) depicts an example of a domain model with an invalid configuration. Using the consistency checking reasoning service and requesting some explanation, the reasoner delivers the answer shown in Listing 1.9:

Listing 1.9. Explanation for domain model inconsistency

```
1 CHECK CONSISTENCY

Consistent: No

5 Explanation:
    Configuration7603 subClassOf Configuration
    config76 type Configuration7603
    card_SPA1 type SPAInterface
    Configuration subClassOf hasSlot min 1 Slot
10                    and hasSlot some hasCard some Supervisor720
    card_SPA2 type SPAInterface
    card_HS type HotSwappableOSM
    Supervisor720 subClassOf Card
```

The reason for the inconsistency here is the missing supervisor card, which must be part of every configuration. Since config76 has as type Configuration7603 which is a subclass of Configuration it must be connected via a slot with some card of type Supervisor720. This is not fulfilled because all cards (card_SPA1, card_SPA2, card_HS) plugged into the configuration are either of type HotSwappableOSM or type SPAInterface.

Classification of Device Configurations. Language users often start modelling with general concepts, e.g. with model elements of type Device or Configuration, depicted in Figure 11 (1). To classify a device or configuration, we have to again transform the instances together with its links into a description logics ABox. The TBox is built by the PDDSL metamodel.

Since description logics allow for simultaneously reasoning on the model (TBox) and instance layer (ABox) we are able to use the *classification reasoning service* (introduced in Section 3.2) to compute all possible types of the individual config76 with regard to all other individuals and relations in the ABox. The result is the more specific type Configuration7603 of the individual config76. Again the individual device76 can be classified. The most specific type here is Cisco7603.

6   Ontology Reasoning for Behaviour Modelling Languages

In this section, we demonstrate the representation of behaviour models in OWL and applications of reasoning services in order to provide model management services like the retrieval of process models based on a process description. We apply our approach on process models, represented by UML Activity Diagrams.

In order to use ontology reasoning for process models, a first step is to build a model bridge from process models (software models) in a UML-like representation to an ontology (TBox). The architecture of a model bridge, or in particular of a process model bridge is depicted in Figure 14. The model bridge is defined in the metamodelling layer M2 and is used in layer M1 to transform or integrate model entities on layer M1. In this section, we consider a transformation bridge.

We present our process model bridge that defines a transformation from process models given as UML activity diagrams to on OWL ontology (TBox). This requires a thorough consideration of the entities that are represented in process models, their relations like control flow relations and how they are transformed to OWL ontologies. A challenge in this task is to capture the semantics of process models like activity ordering and flow conditions in the ontology.

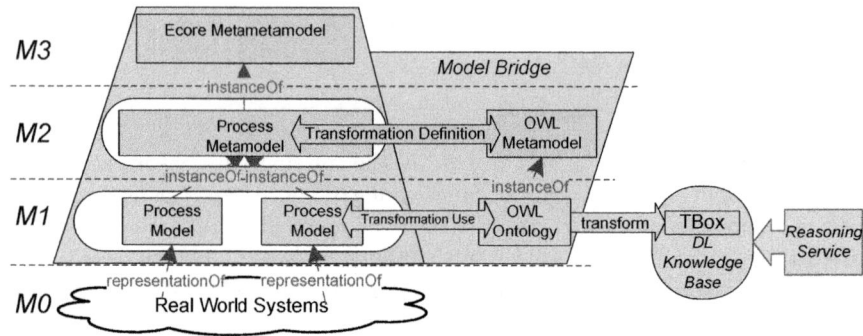

Fig. 14. Process Model Bridge

6.1 Transformation of Process Models into Semantic Representations

Process models capture the dynamic behaviour of an application or system. In software modelling, they are represented by graphical models like BPMN Diagrams or UML Activity Diagrams. The metamodels of both are instances of Ecore meta-metamodels. The two metamodels provide flexible means for describing process models for various applications. However, due to their flexibility further modelling constraints and semantic descriptions are required for a clearer representation of the intended meaning.

We have identified the following additional modelling characteristics for process models in the software modelling space which are analysed in detail in [42]. (1) A semantic representation of control flow dependencies of activities within a process, i.e. execution ordering of activities in a control flow. Such constraints allow for the description of order dependencies e.g., an activity requires a certain activity as a predecessor or successor. (2) It is quite common in model-driven engineering to specialise or refine a model into a more fine-grained representation that is closer to the concrete implementation (cf. [43]). In process modelling, activities could be replaced by sub-activities for a more precise description of a process. Hence, modelling possibilities for sub-activities and also for control flow constraints of these sub-activities should be supported. (3) Quite often, one may formulate process properties that cover modality, i.e. to express a certain property like the occurrence of an activity within a control flow is optional or unavoidable in all possible process instances (traces).

The next subsection gives an overview of the model bridge from a UML activity diagram to an OWL ontology including a discussion of design decisions. The transformation bridge describes how a process model on modelling layer M1 is transformed from a UML representation to an ontology. We demonstrate how entities of an activity diagram like activities, gateways and conditions are modelled in an OWL ontology. Furthermore, we demonstrate the usage of logical representation of a process model in order to achieve the afore-mentioned modelling possibilities in software process modelling. We use the OWL representation of process models in combination with reasoning services to retrieve processes and check process constraints in the second part of this section.

Mapping Process Models to OWL. A process model describes the set of all legal process runs or traces. Activities are represented by OWL classes and a process is modelled as a complex expression that captures all activities of the process. A process run is an instance of this complex class expression in OWL. The process models are described in OWL DL, as syntax we use the DL notation. Basic modelling principles and transformation patterns from UML activity diagrams to OWL are given in Table 2.

Control flow relations between activities are represented by object properties in OWL, i.e. by the object property TO_i (Table 2, No. 4). In order to allow

Table 2. Transformation to OWL

Construct	UML Notation	DL Notation
1. Start	●	$Start_i$
2. End	◉	End_i
3. Activity	Receive Order	$Receive\ Order$
4. Edge	→	TO_i
5. Process P	●→ Receive Order →◉	$P \equiv Start_i \sqcap \exists_{=1} TO_i.$ $(ReceiveOrder \sqcap \exists_{=1} TO_i.End_i)$
6. Flow	Receive Order → Fill Order	$ReceiveOrder \sqcap \exists_{=1} TO_i.FillOrder$
7. Decision	Reject Order / Receive Order → Fill Order → Close Order	$ReceiveOrder \sqcap \exists_{=1} TO_i.$ $((RejectOrder \sqcup FillOrder)$ $\sqcap \exists_{=1} TO_i.CloseOrder)$
8. Condition	[Order accepted] Receive Order → Fill Order	$ReceiveOrder \sqcap \exists_{=1} TO_i.$ $((FillOrder \sqcap \kappa_{OrderAccepted}) \sqcup$ $(Stalled \sqcap \neg\kappa_{OrderAccepted}))$
9. Fork and Join	Ship Order / Receive Order → Send Invoice → Close Order	$ReceiveOrder \sqcap \exists TO_i.$ $(ShipOrder \sqcap \exists_{=1} TO_i.CloseOrder)$ $\sqcap \exists TO_i.(SendInvoice \sqcap$ $\exists_{=1} TO_i.CloseOrder) \sqcap = 2\ TO_i$
10. Loop	Receive Order → Fill Order	$Loop_j \sqcap \exists_{=1}TO_i.FillOrder,$ $Loop_j \equiv ReceiveOrder \sqcap \exists_{=1} TO_j.$ $(Loop_j \sqcup End_j)$

for process composition and refinement in combination with cardinality restrictions, the roles for each process (TO_i) are distinguished from each other. All roles TO_i are defined as sub-roles of TO in order to simplify the retrieval. The object property TO and therefore also the sub-properties TO_i are sub-properties of the transitive property TOT. The property TOT is used to express transitive connections of activities to their predecessor and successor activities with an arbitrary number of activities between them. This kind of hierarchical structuring of object properties is similar to the best practice modelling style for relation pattern as in the DOLCE plan extension [44].

A process is composed by activities and is described in OWL by axioms as shown in No. 5. An axiom defines a process as one complex DL expression capturing all activities that occur in the process. It starts with $Start_i$ followed by a sequence of composed activities. The last activity is the concept End_i.

The control flow (No. 6) is a class expression in OWL like $ReceiveOrder \sqcap \exists TO_i.FillOrder$ meaning the activity $ReceiveOrder$ is directly followed by the activity $FillOrder$. We use concept union for decisions (No. 7). The non-deterministic choice between the activity $RejectOrder$ and $FillOrder$ is given by the class expression $\exists TO_i.(RejectOrder \sqcup FillOrder)$. The activity $CloseOrder$ merges the flows to one outgoing sequence flow. Deterministic choices and exclusive decisions are represented by using different (disjoint) flow conditions.

Flow conditions (No. 8) are assigned to the control flow. The semantics of a flow condition is a restriction on all instances that satisfy the target activity of this flow. Hence, the target instances of the flow are instances of the activity $FillOrder$ and they have to satisfy the $OrderAccepted$ condition, i.e. they are also instances of the class $\kappa_{OrderAccepted}$.

A loop (No. 10) is a special kind of decision. An additional OWL class $Loop_j$ for the subprocess with the loop is introduced to describe multiple occurrences of the activities within the loop. Parallel executions are represented by intersections (No. 9). It is an explicit statement that an activity have multiple successors simultaneously.

6.2   Ontology Reasoning for Process Retrieval

In the previous subsection, we used the expressive power of OWL to provide a semantic representation of process models. Based on this representation, we describe queries in DL and use reasoning services to retrieve processes and process information.

Queries are general and incomplete process descriptions that specify the core functionality of a process of interest. The query result contains all processes of the knowledge base or process repository that satisfy the query specification.

The following example demonstrates a query for process that executes the activity $FillOrder$ before $MakePayment$ with an arbitrary number of activities between them:

$$Q \equiv \exists TOT.(FillOrder \sqcap \exists TOT.MakePayment)$$

To exploit the enriched semantic process representation as presented in the previous subsection, the process retrieval covers three non-disjoint patterns of process structuring and control flow information. (i) A query describes the relevant ordering conditions like which activity has to follow (directly or indirectly) another activity. The transitive object property TOT is used to indicate the possibly indirect connection of the activities. (ii) Besides ordering constraints, semantic query processing allows the retrieval of processes that contain specialised or refined activities, i.e. the process retrieval takes into account the terminological knowledge of the processes like hierarchical structuring of activities. For instance, the result of the demonstrated query also contains all processes that contains sub-activities of $FillOrder$ and $MakePayment$, satisfying the ordering condition. The corresponding class expressions in the OWL model are specialisations of the class expression given by the query expression. (iii) Finally, the usage of the queries allows handling of modality for activity occurrences in a process, like a query that expresses whether the activity $MakePayment$ has to occur in each process or might occur only in some process.

In general, there are various capabilities of DL reasoning or entailment regimes in order to find process models that satisfy or match the process description given by the DL query expression. Again, the query description is also a (general) description of a process. For a more comprehensive overview of different DL inference or entailment methodologies, we refer the reader to [45]. Here, we present process retrieval for two inference methodologies: concept subsumption and concept satisfiability.

Retrieval by Concept Subsumption. In this paragraph, we use entailment of concept subsumption in order to retrieve process models. The process description given by the query is a general and abstract process model. The more complex DL expressions that represent processes in the knowledge base (KB) are more specific with respect to the representation in OWL, i.e. containing further activities, intersections of further parts of the process and (additional) conditions. Hence, all retrieved processes that satisfy the process description of the query are specialisations in OWL of the general process description given by the query, i.e. these processes are subsumed by the more abstract query process.

As an example, we consider a query that searches for all processes that execute the $MakePayment$ activity and its direct successor is the activity $AcceptPayment$. This query (Q) is described by the following DL expression $Q \equiv \exists TOT.(MakePayment \sqcap \exists TO.AcceptPayment)$. The direct successor relation of $MakePayment$ and $AcceptPayment$ is expressed by the non-transitive property TO. Consider the following process that satisfy this query description which is a fragment of the process depicted in Figure 5. $OrderProcess \equiv Start_1 \sqcap \exists_{=1}TO_1.(FillOrder \sqcap \exists TO_1.(ShipOrder \sqcap \exists_{=1}TO_1.End_1) \sqcap \exists TO_1. (MakePayment \sqcap \exists_{=1}TO_1.(AcceptPayment \sqcap \exists_{=1}TO_1.End_1))\sqcap = 2TO_1)$

The knowledge base entails that $OrderProcess$ is subsumed by the query process Q: $KB \models OrderProcess \sqsubseteq Q$

Retrieval by Concept Satisfiability. A weaker entailment regime is checking the satisfiability of concept intersection. The concept intersection is the intersection of the query process description Q and a process from the knowledge base (KB), like the *OrderProcess*. The result of such a query Q are all process models of the knowledge base for which the intersection with the query is a satisfiable concept.

Obviously, the result set is in general larger that in the previous, stronger entailment methodology. The above mentioned process *OrderProcess* would also satisfy this query Q if there are no further statements like the disjointness of activities in the knowledge base. The result of the query Q has to satisfy the following condition:

$KB \cup \{p : (OrderProcess \sqcap Q)\}$ is satisfiable. The p in the query is an instance, i.e. a process run.

The difference of this weaker inference notion becomes more evident if we are looking for a process that is not in the knowledge base, e.g., a different quantifier is used in order to restrict the successors of an activity to a certain given activity like the following query.

$$Q \equiv \exists TOT.(MakePayment \sqcap \forall TO.AcceptPayment)$$

Here, we are looking for all processes that have the activity *MakePayment* in the control flow and the successor of this activity is restricted to *AcceptPayment*. We assume there is no such restriction for a process in the knowledge base. Hence, the stronger inference would not retrieve any process model. However, this weaker entailment would still retrieve the example process *OrderProcess*, since the intersection ($OrderProcess \sqcap Q$) is satisfiable in the knowledge base KB. Again, we assume that no further restriction like disjointness of activities or conditions of process flows are in the knowledge base.

6.3   Summary

In this section, we demonstrated how to use OWL for modelling of process models in order to provide reasoning services to validate process models and retrieve processes and process information. A prerequisite is a transformation bridge that maps on the M1 layer UML activity diagrams to an OWL ontology. The transformation bridge we provided is quite generic and can be adapted to other process modelling languages like the Business Process Modeling Notation (BPMN).

7   A Toolkit for MDE with Ontology Technologies

The TwoUse Toolkit is modelling environment filling the gap between MDE and ontology technologies. It is an implementation of current OMG and W3C standards for developing ontology-based software models and model-based OWL ontologies. It is a model-driven tool to bridge the gap between Semantic Web and Model Driven Software Development.

The TwoUse Toolkit has two User Profiles: model-driven software developers and OWL ontology engineers. The TwoUse Toolkit provides the following functionality to model-driven software developers:

- Describe classes in UML class diagrams using OWL class descriptions.
- Semantically search for classes, properties and instances in UML class diagrams.
- Model variability in software systems using OWL classes.
- Design business rules using the UML Profile for SWRL.
- Extent software design patterns with OWL class descriptions.
- Make sense of UML class diagrams using inference explanations.
- Write OWL queries using SPARQL, SAIQL or the OWL query language based on the OWL Functional Syntax using the query editor with syntax highlighting.
- Validate refinements on business process models.

To OWL ontology engineers, The TwoUse Toolkit provides the following functionalities:

- Graphically model OWL ontologies and OWL safe rules using OMG UML Profile for OWL and UML Profile for SWRL.
- Graphically model OWL ontologies and OWL Safe Rules using the OWL 2 Graphical Editor.
- Graphically model and store ontology design patterns as templates.
- Write OWL queries using SPARQL, SAIQL or the OWL query language based on the OWL Functional Syntax using the query editor with syntax highlighting.
- Specify and safe OWL ontologies using the OWL 2 functional syntax with syntax highlighting.
- specify OWL ontology APIs using the agogo editor.

We have implemented the TwoUse Toolkit in the Eclipse Platform using the Eclipse Modelling Framework [46] and is available for download on the project website[3].

8   Related Work

Modelling complex systems usually requires different modelling languages. Hence, the need of integrating them is apparent. Because of semantic overlap of languages, where synergies can be realized by defining bridges, ontologies provide the chance of semantic integration of modelling languages [47].

In the following, we group related approaches into three categories: Firstly, we present approaches where structural languages are bridged. Secondly, bridges for behavioural languages are described and finally, related work on OWL modelling for building mappings between models (model versions) is discussed.

---

[3] http://code.google.com/p/twouse/

Among approaches of bridges between ontology languages and structural modelling language, one can use languages like F-Logic or Alloy to formally describe models. In [48], a transformation of UML+OCL to Alloy is proposed to exploit analysis capabilities of the Alloy Analyzer [49]. In [50], a reasoning environment for OWL is presented, where the OWL ontology is transformed to Alloy. Both approaches show how Alloy can be adopted for consistency checking of UML models or OWL ontologies. F-Logic is a further prominent rule language that combines logical formulas with object oriented and frame-based description features. Different works (e.g. [51,52]) have explored the usage of F-Logic to describe configurations of devices or the semantics of MOF models. The integration in the cases cited above is achieved by transforming MOF models into a knowledge representation language (Alloy or F-logic). Thus, the expressiveness available for DSL designers is limited to MOF/OCL. Our approach extends these approaches by enabling language designers to specify class descriptions à la OWL together with MOF/OCL, increasing expressiveness.

There are various approaches that build bridges between behavioural modelling languages and ontologies. Here, we only mention those that are related to process modelling in OWL. The OWL-S process model [53] describes processes in OWL. The process specification language [54,55] allows formal process modelling in an ontology. Process models are represented in OWL in combination with petri nets in [56] and in Description Logics for workflows in [57]. Compared to our demonstrated model, these approaches either lack in an explicit representation of control flow dependencies in combination with terminological information of activities like a hierarchical structuring of activities, or retrieval of processes with respect to control flow information is only weakly supported.

9   Conclusion

In this chapter, we described how ontology technologies can be adopted on model-driven engineering. The main artifacts in MDE are modelling languages and transformations. We have shown, that they can be supported by ontology technologies.

In this chapter, after the introduction, we first started with foundations on model-driven engineering, covering modelling languages, transformations and a short characterisation of challenges in model-driven engineering. We continued the chapter with an overview of the Web Ontology Language, ontological modelling and reasoning services. Here, we took a deeper look at the reasoning services and how they could be used in MDE. We first started with standard reasoning services (like consistency and satisfiability checking, or classification) and querying services. All these services are adopted on an ontology which is a representation of the metamodel of a modelling language and conforming models created by the language user. Besides standard reasoning services, we considered explanation and repairing services, where explanation services can inform about debugging relevant facts and repairing services give advises how to repair inconsistent models. A last ontology-based service is the linking service which is used to combine different modelling languages and its models for reasoning on one single representation of a system since systems are described by many different software languages.

In the subsequent section, we showed how to integrate ontology languages with standard metamodelling languages (like, for example, KM3) for an integrated modelling of metamodels together with the abstract syntax component of a modelling language. The integrated metamodels consist on the one side of usual concepts for classes or references, but on the other side they can contain additional semantic expressions, constraints and axioms which are based on OWL. Later they are extracted for adopting ontology technologies like different reasoning services on modelling languages. We described the integration of software and ontology languages by introducing language and model bridges.

The support of modelling languages by ontology technologies was separated into support for structural modelling languages and support for behavioural modelling languages. In the first case, we mainly considered the Ecore metamodelling language and UML, as well as languages for describing configurations. Reasoning services here allow for reasoning on the structure of models, which are prescribed by the metamodel of a modelling language. Possible services are the consistency checking service or the classification service. In the case of behavioural modelling languages, we considered process modelling languages. Here the ontology language was bridged with the process modelling language to describe the semantics of the language and the behaviour of processes.

Before concluding the chapter, we presented the main functionalities of the TwoUse toolkit which implements most of the approaches presented in this paper, followed by an overview of related work.

References

1. Mellor, S., Clark, A., Futagami, T.: Model-driven development. IEEE software 20(5), 14–18 (2003)
2. Atkinson, C., Kuhne, T.: Model-driven development: a metamodeling foundation. IEEE software 20(5), 36–41 (2003)
3. Motik, B., Patel-Schneider, P.F., Horrocks, I.: OWL 2 Web Ontology Language: Structural Specification and Functional-Style Syntax (October 2009), http://www.w3.org/TR/owl2-syntax/
4. Ebert, J.: Metamodels Taken Seriously: The TGraph Approach. In: Kontogiannis, K., Tjortjis, C., Winter, A. (eds.) 12th European Conference on Software Maintenance and Reengineering, Piscataway, NJ. IEEE Computer Society, Los Alamitos (2008)
5. Bildhauer, D., Riediger, V., Schwarz, H., Strauss, S.: grUML-An UMLbased Modeling Language for TGraphs. To appear in Arbeitsberichte Informatik, Universität Koblenz-Landau (2008)
6. OMG: Meta Object Facility (MOF) Core Specification (January 2006), http://www.omg.org/docs/formal/06-01-01.pdf
7. OMG: UML Infrastructure Specification, v2.1.2. OMG Adopted Specification (2007)

8. Budinsky, F., Brodsky, S., Merks, E.: Eclipse modeling framework. Pearson Education, London (2003)
9. Jouault, F., Bezivin, J.: KM3: a DSL for Metamodel Specification. In: Gorrieri, R., Wehrheim, H. (eds.) FMOODS 2006. LNCS, vol. 4037, pp. 171–185. Springer, Heidelberg (2006)
10. Jouault, F., Kurtev, I.: Transforming Models with ATL. In: Bruel, J.-M. (ed.) MoDELS 2005. LNCS, vol. 3844, pp. 128–138. Springer, Heidelberg (2006)
11. OMG: MOF QVT Final Adopted Specification. Object Modeling Group (June 2005)
12. Kalnins, A., Barzdins, J., Celms, E.: Model transformation language MOLA. Model Driven Architecture, 62–76
13. Wagner, R.: Developing Model Transformations with Fujaba. In: Proc. of the 4th International Fujaba Days, pp. 206–275 (2006)
14. Wende, C.: Ontology Services for Model-Driven Software Development. MOST Project Deliverable (November 2009), http://www.most-project.eu
15. Motik, B., Patel-Schneider, P.F., Grau, B.C.: OWL 2 Web Ontology Language Direct Semantics (October 2009), http://www.w3.org/TR/owl2-direct-semantics
16. Parsia, B., Sirin, E.: Pellet: An OWL DL Reasoner. In: Proc. of the 2004 International Workshop on Description Logics (DL2004). CEUR Workshop Proceedings, vol. 104 (2004)
17. Prud'hommeaux, E., Seaborne, A.: SPARQL query language for RDF (working draft). Technical report, W3C (March 2007)
18. Schneider, M.: SPARQLAS: Writing SPARQL Queries in OWL Syntax. Bachelor thesis, University of Koblenz-Landau, German (2010)
19. Glimm, B., Parsia, B.: SPARQL 1.1 Entailment Regimes. Working draft, W3C (January 26, 2010), http://www.w3.org/TR/sparql11-entailment/
20. Knublauch, H., Fergerson, R., Noy, N., Musen, M.: The Protégé OWL plugin: An open development environment for semantic web applications. LNCS, pp. 229–243. Springer, Heidelberg (2004)
21. Kalyanpur, A., Parsia, B., Sirin, E., Hendler, J.: Debugging unsatisfiable classes in OWL ontologies. Web Semantics: Science, Services and Agents on the World Wide Web 3(4), 268–293 (2005)
22. Kalyanpur, A., Parsia, B., Horridge, M., Sirin, E.: Finding all justifications of OWL DL entailments. In: Aberer, K., Choi, K.-S., Noy, N., Allemang, D., Lee, K.-I., Nixon, L.J.B., Golbeck, J., Mika, P., Maynard, D., Mizoguchi, R., Schreiber, G., Cudré-Mauroux, P. (eds.) ASWC 2007 and ISWC 2007. LNCS, vol. 4825, pp. 267–280. Springer, Heidelberg (2007)
23. Reiter, R.: A theory of diagnosis from first principles. Artificial Intelligence 32(1), 57–95 (1987)
24. Horridge, M., Drummond, N., Goodwin, J., Rector, A., Stevens, R., Wang, H.: The Manchester OWL Syntax. In: OWLED 2006 Second Workshop on OWL Experiences and Directions, Athens, GA, USA (2006)
25. Kalyanpur, A.: Debugging and Repair of OWL Ontologies. PhD thesis, University of Maryland, College Park (2006)
26. Kalyanpur, A., Parsia, B., Sirin, E., Cuenca-Grau, B.: Repairing Unsatisfiable Concepts in OWL Ontologies. The Semantic Web: Research and Applications, 170–184
27. Euzenat, J., Shvaiko, P.: Ontology matching. Springer, Heidelberg (2007)
28. Walter, T., Ebert, J.: Combining DSLs and Ontologies using Metamodel Integration. In: Taha, W.M. (ed.) DSL 2009. LNCS, vol. 5658, pp. 148–169. Springer, Heidelberg (2009)

29. Parreiras, F.S., Walter, T.: Report on the combined metamodel. Deliverable ICT216691/UoKL/WP1-D1.1/D/PU/a1, University of Koblenz-Landau, MOST Project (2008)
30. Iqbal, A., Ureche, O., Hausenblas, M., Tummarello, G.: Ld2sd: Linked data driven software development. In: Proceedings of the 21st International Conference on Software Engineering & Knowledge Engineering (SEKE 2009), Boston, Massachusetts, USA, July 1-3, pp. 240–245. Knowledge Systems Institute Graduate School (2009)
31. OMG: Ontology Definition Metamodel. Object Modeling Group (September 2008)
32. O'Connor, M.J., Shankar, R., Tu, S.W., Nyulas, C., Parrish, D., Musen, M.A., Das, A.K.: Using semantic web technologies for knowledge-driven querying of biomedical data. In: Bellazzi, R., Abu-Hanna, A., Hunter, J. (eds.) AIME 2007. LNCS (LNAI), vol. 4594, pp. 267–276. Springer, Heidelberg (2007)
33. Staab, S., Scherp, A., Arndt, R., Troncy, R., Gregorzek, M., Saathoff, C., Schenk, S., Hardman, L.: Semantic multimedia. In: Baroglio, C., Bonatti, P.A., Małuszyński, J., Marchiori, M., Polleres, A., Schaffert, S. (eds.) RW 2008. LNCS, vol. 5224, pp. 125–170. Springer, Heidelberg (2008)
34. Staab, S., Franz, T., Görlitz, O., Saathoff, C., Schenk, S., Sizov, S.: Lifecycle Knowledge Management: Getting the Semantics Across in X-Media. In: Esposito, F., Raś, Z.W., Malerba, D., Semeraro, G. (eds.) ISMIS 2006. LNCS (LNAI), vol. 4203, pp. 1–10. Springer, Heidelberg (2006)
35. Silva Parreiras, F., Staab, S.: Using ontologies with uml class-based modeling: The twouse approach. Data Knowl. Eng. (to be published)
36. Walter, T., Silva Parreiras, F., Staab, S.: OntoDSL: An Ontology-Based Framework for Domain-Specific Languages. In: Schürr, A., Selic, B. (eds.) MODELS 2009. LNCS, vol. 5795, pp. 408–422. Springer, Heidelberg (2009)
37. Walter, T., Silva Parreiras, F., Staab, S., Ebert, J.: Joint language and domain engineering. In: Kühne, T., Selic, B., Gervais, M.-P., Terrier, F. (eds.) ECMFA 2010. LNCS, vol. 6138, pp. 321–336. Springer, Heidelberg (2010)
38. Gamma, E., Helm, R., Johnson, R., Vlissides, J.: Design patterns: elements of reusable object-oriented software. Addison-Wesley Professional, Reading (1995)
39. Silva Parreiras, F., Staab, S., Schenk, S., Winter, A.: Model driven specification of ontology translations. In: Li, Q., Spaccapietra, S., Yu, E., Olivé, A. (eds.) ER 2008. LNCS, vol. 5231, pp. 484–497. Springer, Heidelberg (2008)
40. Walter, T., Ebert, J.: Combining ontology-enriched domain-specific languages. In: Proceedings of the of the Second Workshop on Transforming and Weaving Ontologies in Model Driven Engineering (TWOMDE) at MoDELS (2009)
41. Miksa, K., Kasztelnik, M., Sabina, P., Walter, T.: Towards semantic modelling of network physical devices. In: Ghosh, S. (ed.) MODELS 2009. LNCS, vol. 6002, pp. 329–343. Springer, Heidelberg (2010)
42. Groener, G., Staab, S.: Modeling and Query Pattern for Process Retrieval in OWL. In: Bernstein, A., Karger, D.R., Heath, T., Feigenbaum, L., Maynard, D., Motta, E., Thirunarayan, K. (eds.) ISWC 2009. LNCS, vol. 5823, pp. 243–259. Springer, Heidelberg (2009)
43. Ren, Y., Groener, G., Lemcke, J., Rahmani, T., Friesen, A., Zhao, Y., Pan, J.Z., Staab, S.: Validating Process Refinement with Ontologies. In: International Workshop on Description Logics (2009)
44. Gangemi, A., Borgo, S., Catenacci, C., Lehmenn, J.: Task Taxonomies for Knowledge Content D07. In: Metokis Project public Deliverable, pp. 20–42 (2004)
45. Grimm, S., Motik, B., Preist, C.: Variance in e-Business Service Discovery. In: Proc. of the ISWC Workshop on Semantic Web Services (2004)

46. Budinsky, F., Brodsky, S.A., Merks, E.: Eclipse Modeling Framework. Pearson Education, London (2003)
47. Gasevic, D., Djuric, D., Devedzic, V.: Model Driven Architecture and Ontology Development. Springer, Heidelberg (2006)
48. Anastasakis, K., Bordbar, B., Georg, G., Ray, I.: UML2Alloy: A challenging model transformation. In: Engels, G., Opdyke, B., Schmidt, D.C., Weil, F. (eds.) MODELS 2007. LNCS, vol. 4735, pp. 436–450. Springer, Heidelberg (2007)
49. Jackson, D.: Software Abstractions: logic, language, and analysis. The MIT Press, Cambridge (2006)
50. Wang, H., Dong, J., Sun, J., Sun, J.: Reasoning support for Semantic Web ontology family languages using Alloy. Multiagent and Grid Systems 2(4), 455–471 (2006)
51. Sure, Y., Angele, J., Staab, S.: OntoEdit: Guiding ontology development by methodology and inferencing. LNCS, pp. 1205–1222. Springer, Heidelberg
52. Gerber, A., Lawley, M., Raymond, K., Steel, J., Wood, A.: Using Sophisticated Models in Resolution Theorem Proving. LNCS, pp. 90–105. Springer, Heidelberg (2002)
53. Martin, D.: OWL-S: Semantic Markup for Web Services (2004), http://www.w3.org/Submission/OWL-S
54. Grüninger, M., Menzel, C.: The Process Specification Language (PSL) Theory and Application. AI Magazine 24, 63–74 (2003)
55. Grüninger, M.: 29. In: Ontology of the Process Specification Language, pp. 575–592. Springer, Heidelberg (2009)
56. Koschmider, A., Oberweis, A.: Ontology Based Business Process Description. In: EMOI-INTEROP (2005)
57. Goderis, A., Sattler, U., Goble, C.: Applying DLs to workflow reuse and repurposing. In: Description Logic Workshop (2005)

Combining Ontologies with Domain Specific Languages: A Case Study from Network Configuration Software*

Krzysztof Miksa, Pawel Sabina, and Marek Kasztelnik

Comarch SA
Al. Jana Pawla II 39A, Krakow, Poland
{krzysztof.miksa,pawel.sabina,marek.kasztelnik}@comarch.com

Abstract. One of the important aspects of Model-Driven Engineering (MDE) is to consider application-domain variability, which leads to creation of Domain Specific Languages (DSL). As with DSLs models are concise, easy to understand and maintain, this approach greatly increases the productivity and software quality. Usually, the DSLs in MDE are described with a metamodel and a concrete syntax definition. The models expressed in the DSL are linguistic instantiations of the language concepts found in the metamodel.

However, for some of the application domains it's not enough to consider the linguistic dimension of the instantiation. The problem arises when the domain itself contains the aspect of typing. This leads to another view on instantiation, called *ontological instantiation* . Since both aspects are present simultaneously, we refer to the combined approach with the term "two-dimensional metamodelling".

In the following, we will exemplify the problem with a case study based on a real challenge found in the domain of network management. The solution we propose benefits from ontology technology which is applied to enforce the semantics of ontological instantiation. Our approach presents significant differences comparing to the existing 2D metamodelling solution, although the motivations are similar. Thus, we consider our work as a case study of applying ontology enabled software engineering in the area of DSL engineering, rather than a new metamodelling technology or an application of existing 2D metamodelling architecture.

The article is a result of joint work of the MOST project partners, applied within the case study provided by Comarch.

1   Introduction

Facing global competition, the software companies are striving to respond to customer's need for flexible, large scale, software systems that can be easily

---

* This research has been co-funded by the European Commission and by the Swiss Federal Office for Education and Science within the 7th Framework Programme project MOST number 216691.

U. Aßmann, A. Bartho, and C. Wende (Eds.): Reasoning Web 2010, LNCS 6325, pp. 99–118, 2010.

and quickly customized, configured and deployed. In large software development organizations, increased complexity of products, shortened development cycles, and heightened expectations of quality have created major challenges at all the stages of the software lifecycle. To respond to these requirements, various software-engineering methodologies have been used to capture the information concerning requirements, architecture, design, implementation and testing. Many of these methodologies promote usage of models to capture and maintain that information. A prominent example is the conceptualization of business entities that exist in system domain in class diagrams available in the Unified Modeling Language (UML).

1.1   Model-Driven Engineering and Domain Specific Languages

Model-Driven Engineering (MDE) [1] combines three main ingredients: 1) an adequate abstraction for specifying different system aspects with 2) a methodology to automate the transformation of these specifications into a system implementation, and 3) an explicitly defined process describing the application of MDSD [2]. At Comarch [1], we adopt a practical approach of combining both model-driven and traditional software development. This is due to the fact that the full transition of large, already existing product lines to model-driven paradigm is very difficult process. Despite this, model-driven approaches constantly gain more attention.

One of the reasons for this is the productivity of Domain Specific Languages, which alleviate the complexity of the software systems and allow for expressing domain concepts effectively   [3]. Domain Specific Languages (DSLs) provide a simplified description of a specific domain. Maintaining models using small but clearly defined concepts and syntax increases productivity, resistance to mistakes and readability of the designed models. The models are typically used to generate code, possibly through a chain of transformations, or interpreted directly.

1.2   Shortcomings of DSL Engineering

However, although DSLs constitute a significant step forward in software development, there are still a number challenges. Firstly, DSLs still should be expressive enough to represent a broad range of possible cases found in the domain. Thus, a trade-off exists between productivity and expressiveness. A complicated DSL leads to lower productivity, while limiting expressive power may lead to the oversimplification and, in consequence, limited use of the DSL, since complicated cases need to be solved with other measures - typically in code.

The infrastructure of the DSL comprises several artefacts. This obviously includes elements such as the abstract and concrete syntax specifications, the well-formedness rules, etc. Additionally, in order to enable the application and to increase productivity, various additional tooling is also provided. Typically, these tools are highly dependent on the semantics of the language constructs. However, current metamodelling technologies lack the means to specify semantics. Thus, in order to provide a tool infrastructure and to guide the developer in modelling,

---

[1] `www.comarch.com`

the semantics of the language is hard coded in the tools. Therefore, there is a clear need for a formalisms and techniques to provide an advanced support in order to help the users to make correct decisions. This can be provided by the tools based on knowledge and semantics, able to reason and bring meaningful guidance and answers for the user questions.

1.3   Ontologies to Improve DSL Engineering

Ontologies can be used to specify semantics and reasoners can be employed to check consistency of models and to infer implicitly modelled information [2]. Therefore, ontologies can be the answer to the shortcomings of MDE and DSL engineering in particular. The semantics of language can be specified in an ontology, and the DSL tools can benefit from the semantic reasoning services to prevent hard coding the semantics. Additionally, by combining the concepts from classical models and the ontology concepts it is possible to extend the expressiveness of DSL, while keeping it simple and productive.

However, the ontologies and models constitute different technological spaces that must be bridged. Our recent research in the MOST project revealed that there are various options for bridging and configuring the reasoning services. In the following, after introducing the problem domain, we will describe the specific option chosen for the case of modelling physical devices. The bridge in this case is based on both metamodel integration and transformation.

1.4   Structure of the Paper

The paper is organized as follows: in the next section we introduce the case study of modelling physical network devices. Section 3 describes the work that we have performed in order to bridge the modelling and ontology technical spaces. In Section 4, we describe the ontological solution for modelling physical devices. Section 5 describes the prototype application which was created to evaluate our approach. The section contains also examples which show how the reasoning services are applied in the specific scenarios of using the prototype. Finally, we discuss the related work in Section 6 and summarize our key findings in Section 7.

2   Case Study: Modelling of the Physical Network Devices

One of the challenges faced in the domain of Next Generation Operation Support Systems [4] is the increasing need for the consistent management of the physical network equipment. However, the state-of-the-art solutions are unable to provide any consistent support to the user by answering questions that involve the sophisticated, configuration related constraints. Myriads of device types and their configurations can make user's everyday work a nightmare.

Let us take an example of a usual situation in the telecommunication companies when one of the physical device cards is broken and requires replacement. The process of finding a valid replacement requires deep knowledge about every sub-component of the physical device (what kind of cards can be used as a

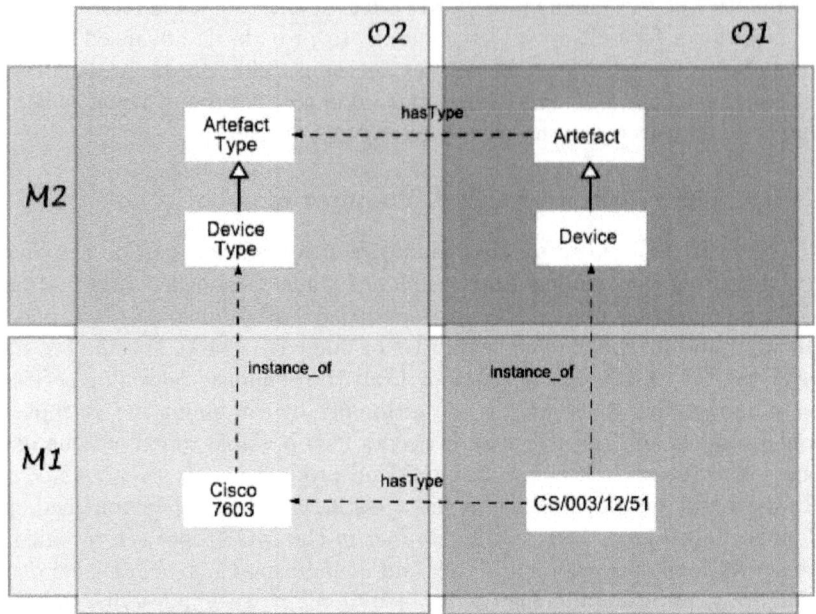

Fig. 1. Two dimensional metamodelling in the PDDSL case

replacement for a broken card, what kind of constraints a particular card has, etc.). In the state-of-the-art solutions, this problem is either completely ignored, leaving it to the expertise of the user, or the rules are hard coded in the solution.

2.1   Physical Devices Domain Specific Language

At Comarch, we use a DSL that describes the structure of the physical device and stores the information about the possible connections between physical devices, as well as the representation of the concrete instances of the devices: the Physical Devices DSL (PDDSL). The language contains both the concepts that refer to the types of the devices (e.g. the Cisco 7603 router model), as well as the instances of these types (e.g. a concrete Cisco 7603 router operating in the network).

Therefore, our approach relates to the *two-dimensional metamodelling architecture* [5], with both the *linguistic* and the *ontological* instantiation. The idea of having two different types of instantiation is given in Figure 1.

The diagram consists of the two *linguistic* layers:

- The M2 layer contains the PDDSL metamodel.
- The M1 layer contains the PDDSL model.

Also, the two *ontological* layers are outlined:

- The O2 layer describes the types of the physical devices.
- The O1 layer describes the instances of the physical devices.

The instance_of relationship between the elements of the layers M1 and M2 is the *linguistic* instantiation relationship (specifies a linguistic type of the modelling object), e.g. Cisco7603 is the instance of the DeviceType metaconcept. The hasType relationship between the elements of the layers O1 and O2 existing at the M1 are the occurrences *ontological* instantiation relationship (specifies the domain type of the modelling object), e.g. the Cisco7603 is the domain type of the 'CS/003/12/51' device.

The similar hasType association at the M2 level represents the *linguistic* definition of the *ontological* instantiation relationship. It connects two classes, the Artefact and the ArtefactType, at the M2. These two classes are fundamental for the PDDSL metamodel, which is depicted in Figure 2. The Artefact is a common supertype for all classes that constitute the O1 layer, such as Slot, Card and Device. The ArtefactType is a common supertype for all classes that constitute the O2 layer, such as SlotType, CardType and DeviceType.

For exemplification, let's consider the simplest type of device found in the Cisco 7600 family, the Cisco 7603 router (Listing 1), which contains three slots. The first slot is reserved for a supervisor card, which is required. The second slot can optionally contain a backup supervisor or a Catalyst_6500_Module. The third slot is also optional and can contain only a Catalyst_6500_Module.

```
DeviceType "Cisco_7603"
longName : "CISCO 7603 CHASSIS"
  allowed : {
    PossibleConfiguration "Cisco_7603_Configuration" {
      SlotType "1" allowed : "Supervisor_Engine_2" "Supervisor_Engine_720" required : true
      SlotType "2" allowed : "Supervisor_Engine_2" "Supervisor_Engine_720"
          "Catalyst_6500_Module" required : false
      SlotType "3" allowed : "Catalyst_6500_Module" required : false
    }
  }
```

Listing 1. Cisco 7603 type model

As an example of instance model (M1, O1), let's consider an instance of the device type Cisco_7603 (Listing 2). The model is invalid since the slot "1" contains a card which is disallowed in this slot.

```
Device serialNumber : "cisco_7603" hasType : "Cisco_7603"
    configuration : {
        Slot id : "1" :
            Card serialNumber : "sip−400" hasType : "7600−SIP−400"
        Slot id : "2" :
        Slot id : "3" :
    }
```

Listing 2. Sample incorrect instance model of Cisco 7603

The reason why we define the `hasType` relationship explicitly in the meta-model is that our case does not fit into a single type/instance level architecture. However, despite that the metamodel definition contains the `hasType` relationship, such an association has no specific meaning to the modelling tools. For instance, the model elements located in the layers M1, O2 have no role in determining whether their instances in the layers M1, O1 are well formed [6].

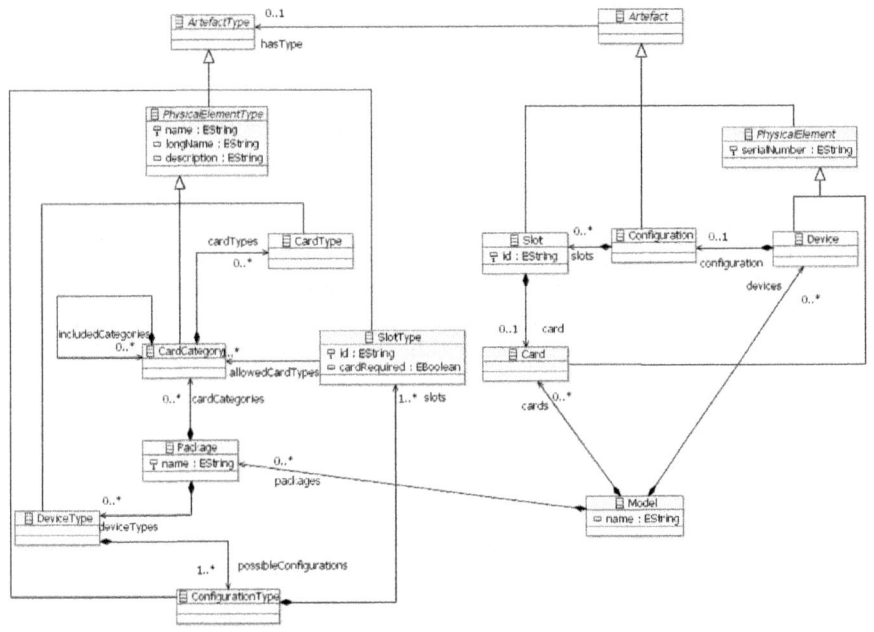

Fig. 2. Simplified version of the Physical Devices DSL metamodel (in Eclipce Ecore)

The language was engineered using the Eclipse Modelling Framework [7]. The PDDSL has a simple textual concrete syntax that enables the textual definition of models. The sample models expressed in this syntax with some explanation can be found in Section 5.1. The syntax for the language was provided using the EMFText [8] tool.

2.2   Services Required by PDDSL Users

The capabilities of the DSL tools are motivated by the tasks which are performed by the language users. The users of the language fall into the two categories (roles):

- **The device expert** interacts with the type layer of the PDDSL and defines the configuration constraints of the devices. Device expert is an expert in the domain of network devices and possesses skills in ontologies.

- **The device modeller** interacts with the instance layer of the PDDSL and models the concrete installations of the devices. Device modeller des not have any skills in ontologies.

In the following, the three typical use cases (UC) of the PDDSL users are formalized in a form of the use case descriptions.

UC-1: Detect errors in physical device type definition

Goal	Support the Device Expert by validating the physical devices type definitions.
Actor	Device Expert
Pre-Condition	An initial physical devices type definition.
Trigger of Use Case	After a change in the physical devices types model, the Device Expert invokes the validation service.
Post-Condition	The information about the successful validation or the error. The error message includes the reason why the model is incorrect and the model element which are wrong.

UC-2: Find wrongly configured instances of devices and explain errors

Goal	Identify wrong device instances, and explain the reasons.
Actor	Device Modeller
Pre-Condition	A physical devices type definition. An initial instance model.
Trigger of Use Case	After creating a device configuration, the user wants to check its correctness.
Post-Condition	The information about the successful validation or the error. The error message includes the reason why the model is incorrect and the model element which are wrong.

UC-3: Suggest card categories which are allowed in a slot

Goal	Support the user by providing suggestions about the possible configurations. The suggestion enumerates the card categories that can be plugged into a slot in a given configuration instance.
Actor	Device Modeller
Pre-Condition	The physical devices type definition. The initial instance model with at least one slot.
Trigger of Use Case	Device Modeller configures the device instance and asks for guidance.
Post-Condition	The list of the card categories that are allowed in the slot.

3   Integrating Models and Ontologies

The existing works that compare the ontological and metamodelling technology spaces [9] reveal that the hybrid approaches present some significant advantages over the technologies stemming purely from the metamodelling space, such as the OCL. Our work goes toward extending the expressiveness of the domain specific language for the description of the physical device structure. This is achieved through the integration of the Domain Specific Languages with ontologies, and thus opening the new spaces for modelling the structure and the semantics together. The integrated models remain easy to edit, process, and reasoned about using the existing tools and approaches, being a structural and semantic model at the same time.

To achieve these goals, we follow an approach proposed by our partner in the MOST project, the University of Koblenz-Landau[2], based on the idea of integration of the models and ontologies [10]. The bridging between the modelling technical spaces and the OWL2 technical space is based on both the integration of the languages (PDDSL and OWL2) and the transformation technology. The language integration is the more fundamental part, as the transformation already assumes that the languages are integrated.

In this chapter, we describe how OWL2 can be represented in the metamodelling technical spaces and describe the integration between OWL2 and PDDSL. The integration takes place at the abstract syntax (metamodel) level and the concrete syntax level.

3.1   OWL2 Manchester Syntax Metamodel

We use the *OWL2 Manchester Syntax* [11] to represent the OWL2 ontology. The corresponding metamodel is large and complicated, since it reflects the complete specification of the Manchester Syntax. Figure 3 shows a small excerpt of the metamodel with some of the core concepts. The ontology in the Manchester Syntax consists of the frames representing, among other entities, the classes and the individuals. The definition of a class can contain the descriptions, representing the related class expression. For instance, through `equivalentClassesDescriptions` it is possible to define the equivalent classes. Similarly, an individual can be related to the class description representing its type, through the `types` reference. The metamodel and the corresponding concrete syntax is available at the EMFText Concrete Syntax Zoo[3].

3.2   Metamodel Integration

The integration of the OWL2 and PDDSL metamodels is depicted in Figure 4. Firstly, we state that the `ArtefactType` class from the PDDSL metamodel is a subclass of the `Class` from OWL2 metamodel. Secondly, we define the `Artefact`

---

[2] http://isweb.uni-koblenz.de

[3] http://www.reuseware.org/index.php/EMFText_Concrete_Syntax_Zoo

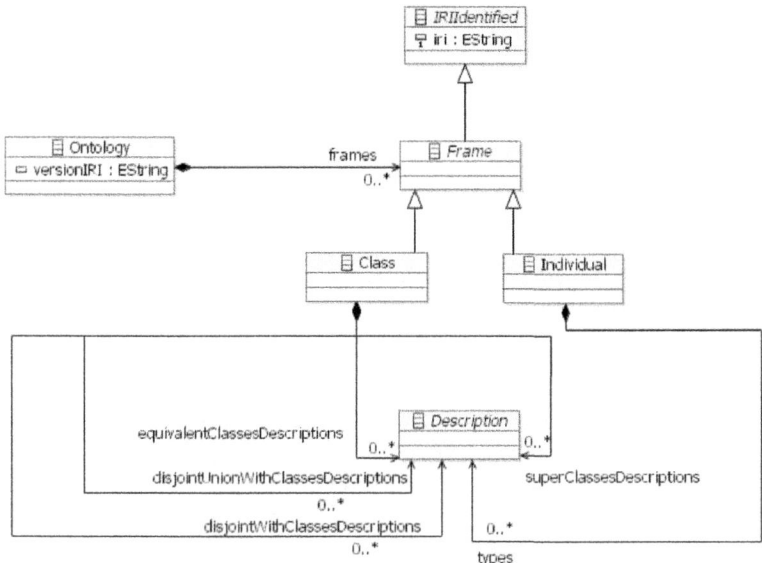

Fig. 3. OWL2 Manchester Syntax metamodel excerpt

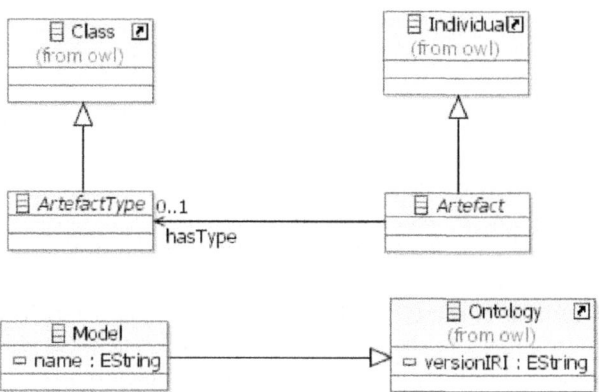

Fig. 4. PDDSL-OWL2 metamodel integration

from PDDSL metamodel as a subclass of the Individual from OWL2 meta-model. Finally, we state that the Model from the PDDSL metamodel is a subclass of the Ontology from the OWL2 metamodel.

3.3   Concrete Textual Syntax Integration

The textual syntax specifications of PDDSL and OWL2 were integrated by altering the PDDSL syntax, so that the enrichment of PDDSL constructs with the OWL2 descriptions is possible. For instance, the PDDSL constructs that are subclasses of the Class from the OWL2 metamodel, can be attributed with the OWL2 descriptions which are allowed for the classes (e.g. the equivalent classes description), as well as the annotations (Listing 3 contains an example for the CardCategory).

```
CardCategory::= !1"CardCategory" iri[","]
  (
    (annotations !1)
      | ("SubClassOf:" superClassesDescriptions
          ("," superClassesDescriptions )* !1)
      | ("EquivalentTo:" equivalentClassesDescriptions
          ("," equivalentClassesDescriptions)* !1)
      | ("DisjointWith:" disjointWithClassesDescriptions
          ("," disjointWithClassesDescriptions )* !1)
      | ("DisjointUnionOf:" disjointUnionWithClassesDescriptions
          ("," disjointUnionWithClassesDescriptions)* !1)
  )*
  ("longName"  ":" longName[","])? ("description"  ":" description[","])?
          (!1"cardTypes"  ":" "{" cardTypes* !0"}")?
          (!1"includedCategories"  ":" "{"includedCategories* !0"}" )?   ;
```

Listing 3. Integrated syntax definition for CardCategory (in EMFText concrete syntax specification language)

3.4   Transformation

The purpose of the model transformation, specified in the QVT Operational [12] language, is to convert an input PDDSL model into a pure OWL2 ontology. The transformation step is needed despite the fact that the PDDSL is integrated with OWL2 and can be treated as an ontology metamodel. The reason is that such integrated models cannot be directly fed into the semantic reasoner.

First, the transformation reorganises the structure of the model. The OWL2 Manchester syntax is a frame-based language, while the PDDSL uses a custom structure of containment of the model elements. The elements corresponding to the OWL2 classes or individuals have to be moved into the top level collection of frames.

```
// Move all OWL::Class objects
model.frames += model.allSubobjects() [OWL::Class] -> sortedBy(iri);
```

Listing 4. QVTO - Example of reorganizing PDDSL model elements

Besides moving the objects from one collection to another, the information implicitly defined in the PDDSL has to be explicitly formalized in the ontology. For instance, the class types listed as the value of the SlotType.allowedCardTypes property form the ObjectPropertyOnly restriction on the hasCard property (Listing 5).

```
mapping inout SlotType::updateSlotType() {

    // Define a specific Slot subclass containing
    // only allowed CardTypes, like in example:
    //
    // pd_hasCard only (Supervisor_Engine_2
    //              or Supervisor_Engine_720)
    self.superClassesDescriptions += object ObjectPropertyOnly {
        featureReference := object FeatureReference {
            feature := hasCardProperty;
        };
        primary := object NestedDescription {
            description := object Disjunction {
                self.allowedCardTypes -> forEach (i) {
                    conjunctions += object ClassAtomic{
                        clazz := i;
                    }
                }
            }
        }
    };
}
```

Listing 5. QVTO - Example of processing SlotTypes

As an example, let's consider the output model presented in Listing 6. After the execution of the mapping for `SlotType` "1", the `SlotType` is enriched with a `SubClassOf` axiom. The axiom is based on a universal restriction on the `pd:hasCard` object property. The card categories listed in the restriction correspond to the card categories specified in the source model (Listing 1) for the `SlotType` "1".

```
DeviceType "Cisco_7603"
longName : "CISCO 7603 CHASSIS"
  allowed : {
    PossibleConfiguration "Cisco_7603_Configuration" {
      SlotType "1"
            SubClassOf:
          pd:hasCard only ( Supervisor_Engine_2 or Supervisor_Engine_720 ) ,
            allowed : "Supervisor_Engine_2" "Supervisor_Engine_720" required : true
      [...]
    }
  }
```

Listing 6. Cisco 7603 type model after exacution of mapping

Furthermore, the transformation defines the semantics of the *ontological instantiation* relationship: the `hasType` property. The semantics of this relationship is equivalent to the class assertion. Thus, every occurrence of the `hasType` is transformed to an appropriate class assertion axiom.

4   Physical Devices Ontology

In this section, we describe the design of the ontological solution for the modelling physical devices. The ontology is the result of the integration and transformation process described in Section 3.

Ontology TBox. The TBox of the physical ontology consists of two parts. Firstly, the core concepts of the ontology, which are independent of the particular device types being modelled, are defined (Listing 7). This includes classes Device, Card, Configuration, Slot and the object and data properties that refer to these classes.

Class: pd:Device

Class: pd:Card

Class: pd:Configuration

Class: pd:Slot

ObjectProperty: pd:hasSlot
  Domain: pd:Configuration
  Range: pd:Slot
  Characteristics: InverseFunctional
  InverseOf: pd:isInConfiguration

ObjectProperty: pd:hasConfiguration
  Domain: pd:Device
  Range: pd:Configuration
  Characteristics: InverseFunctional , Functional

ObjectProperty: pd:hasCard
  Domain: pd:Slot
  Range: pd:Card
  Characteristics: InverseFunctional , Functional
  InverseOf: pd:isInSlot

DataProperty: pd:id
  Domain: pd:Slot
  Characteristics: Functional

Listing 7. Core concepts of the physical device ontology

Secondly, the TBox contains the definitions of the particular types of devices. This part reflects the (M1,O2) part (see Figure 1) of the PDDSL model and includes the configuration constraints of the devices. An example excerpt of this part of TBox, with basic information about the Cisco 7603 router, is given in Listing 8.

```
Class: Cisco_7603
  SubClassOf:  pd:Device, pd:hasConfiguration exactly 1 (Cisco_7603_Configuration)


Class: Cisco_7603_Configuration
  SubClassOf:  pd:Configuration , pd:hasSlot exactly 3  pd:Slot

Class: Cisco_7603_Configuration_slot_1
  SubClassOf:  pd:Slot ,
          pd:hasCard only  ( Supervisor_Engine_2 or  Supervisor_Engine_720 ) ,
          pd:hasCard some  pd:Card
  EquivalentTo: pd_isInConfiguration some Cisco_7603_Configuration and pd:id value 1
```

Listing 8. TBox excerpt describing the Cisco 7603 router

The `Cisco_7603_Configuration_slot_1` class is an example slot type defini-tion. It is equivalent to the slot with id "1" of the `Cisco_7603_Configuration`. It is mandatory to put some card in this kind of slot. Also, the slot type allows only for the following categories of cards: `Supervisor_Engine_2` or `Supervisor_Engine_720`.

Ontology ABox. The ABox of the ontology describes the instance part of the model (M1, O1 in Figure 1). The instances reflect the real devices, their configurations, slots, cards and relations among them.

Closing the world. The reasoning tasks performed on models, such as con-sistency checking, often require the Closed Domain Assumption or even Closed World Assumption [13]. In contrast, OWL2 adopts the Open World Assump-tion. Therefore it is necessary to be able to close the knowledge explicitly, e.g. by adding the respective axioms to the ontology. Similarly, OWL2 doesn't as-sume the Unique Name Assumptions which is required to for the purpose of guidance services.

In our prototype, these problems are solved by adding additional axioms to the ontology. Our experiments revealed that in the physical devices use case it cannot be decided in general, what should be closed in the ontology. Rather, some of the decisions depend on the use case. Specifically, the use case UC-2 (Section 2.2) requires closing the pd:hasCard object property to perform the consistency checking, while the UC-3 (Section 2.2) requires the same property to remain open.

All aspects of domain closure independent of a particular use case are realised in the transformation. An example is the addition of the disjoint classes axioms. It is assumed that all top-level card categories are disjoint and that the subcategories within same `CardCategory` are mutually disjoint. Then, it is enough to close the `pd:Card` class (the root class of the `CardCategory` inheritance hierarchy) in order to ensure the Closed Domain Assumption for all of the branches in the hierarchy.

In contrast, the `pd:hasCard` object property has to be closed for the purpose of the consistency checking, while the same property has to remain open for other services. In order to close the property, for each individual that represents a slot without known cards it is stated that it is of the type `pd:hasCard exactly 0 pd:Card`. After performing the service, the added axioms should be removed from the ontology.

Additionally, we adopt the Unique Name Assumption for all individuals. Listing 9 provides a generic model transformation that makes all individuals in an ontology different.

```
model.frames += object DifferentIndividuals{
    individuals += model.allSubobjects() [OWL::Individual];
};
```

Listing 9. Unique Name Assumption for all individuals

5   Prototype Solution

A prototype implementation of the physical devices modelling tool was developed using the Eclipse Modelling Framework [7]. The goal of the tool is to enable modelling physical devices in the DSL and, at the same time, take advantage of the formal semantics and the expressivity of OWL2. The conceptual architecture of the prototype is depicted in Figure 5. The user interacts with the `Integrated modelling` component which enables creating structural models of the network devices (`PD Modelling`) annotated with the additional configuration constraints expressed in the Web Ontology Language (`OWL Modelling`).

The integrated models are then transformed into the pure semantic descriptions (the `Ontology`) which can be accepted by the `Semantic Reasoner`. The prototype was tested to work with two reasoners: Pellet[4] and TrOWL[5]. TrOWL is in fact a reasoning infrastructure that delegates the reasoning services to the other reasoners. Additionally, it can approximate the ontology into a language which offers the lower computational cost of reasoning. We have successfully experimented with the approximation to OWL2-EL [14].

Then, it is possible to invoke the `Guidance Services` which implement the use cases listed in Section 2.2. These services can then benefit from the reasoning

---

[4] `clarkparsia.com/pellet/`
[5] `trowl.eu`

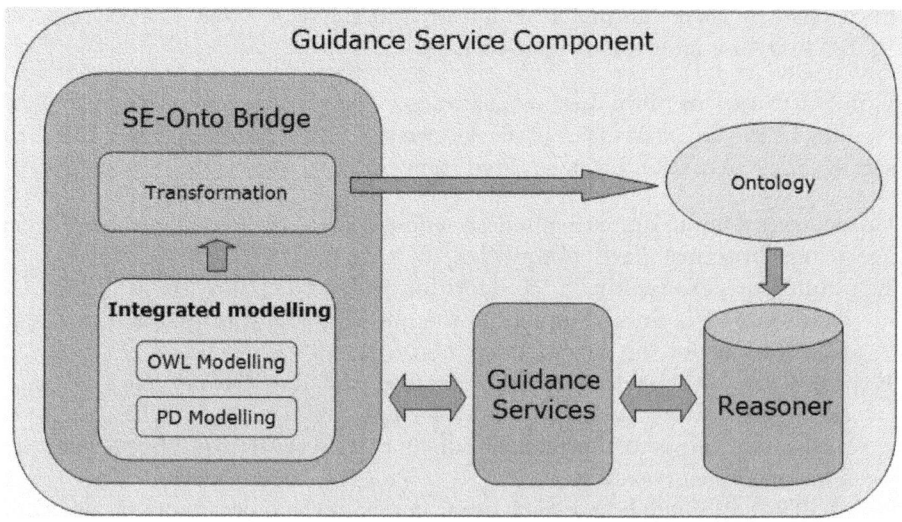

Fig. 5. Physical Devices Guidance Service Component

services provided by the Reasoner, such as the consistency checking, the satisfiability checking and classification. The reasoning results need to be interpreted and returned back to the Integrated modelling environment where they are presented to the user. The interpretation procedure relies on the reasoning explanations.

5.1   Guidance Services

In this section we revisit the use cases defined in Section 2.2. For each of them the corresponding service implementation is described.

Compute incorrect device types. The validation of the DeviceTypes is required by the use case UC-1. The service is based on the satisfiability checking, which is realised by a direct call to the reasoner. For the OWL2 classes contained in an unsatisfiable classes set, their counterpart DeviceTypes are identified and returned to the user as incorrect.

As an example, let's consider the Cisco 7603 router model (Listing 1) and let's suppose that the following additional OWL2 axiom is added to the model: Cisco_7603_Configuration SubClassOf: pd:hasSlot some (pd:hasCard some Cisco_7600_SIP). The axiom requires the only possible configuration of the Cisco_7600 to contain at least one Cisco_7600_SIP card. The constraint is contradictory to the PDDSL specification (Listing 1) which does not allow the Cisco_7600_SIP in any of the slots.

The problem can be detected after the transformation of the integrated PDDSL+OWL2 model to the pure OWL2 ontology. The reasoner will infer that the OWL2 class corresponding to the Cisco_7603 device type is unsatisfiable.

Such result is then interpreted as an error in the type layer of the integrated PDDSL+OWL2 model and reported back to the user.

Compute and explain inconsistencies. The consistency checking and explaining of the inconsistencies is the service required by the use case UC-2. The service implementation includes three steps:

Consistency checking is realised via a direct call to the reasoner which returns answer in terms of a Boolean value.

Explanation generation is an algorithm which computes the Minimum Inconsistency Preserving Subsets of the ontology [15] (also referred to as the justifications for the inconsistency).

Explanation interpretation The reasoning explanation is not itself meaningful for the Device Modeller. Therefore, it is post-processed. The aim of the post-processing is to interpret the explanation in the domain specific manner. This is done in two steps:

1. The set of individuals that occur in any of the object property assertion axioms involved in the explanation is extracted. These individuals are reported to the user as invalid.
2. In order to report the user the reason for the inconsistency, we rely on the annotation mechanism of OWL2. The assumption is that every axiom which is meaningful to the user is annotated with a user friendly description of the error. Such description is then reported to the user if such axiom occurs in the explanation.

Let's consider the instance of the device type Cisco_7603 from Listing 2 and the respective type model presented in Listing 1. The inconsistency detected by the reasoner is related to the axioms describing the slot "1" (Listing 10). Specifically, the only restriction on pd:hasCard property lists the allowed card categories in the slot. This axiom appears in the inconsistency explanation. The OWL2 definition of the slot, generated from the PDDSL model contains also the annotation: "Slot requires card from the following card categories: Supervisor_Engine_2, Supervisor_Engine_720.". This description is interpreted as the error message presented to the user.

Class: Cisco_7603_Configuration_slot_1
  Annotations: rdfs:comment
    "Slot requires card from the following card categories:
    Supervisor_Engine_2, Supervisor_Engine_720."
  SubClassOf: pd:Slot ,
          pd:hasCard only (Supervisor_Engine_2 or Supervisor_Engine_720) ,
          pd:hasCard some pd:Card
  EquivalentTo: pd:isInConfiguration some Cisco_7603_Configuration and pd:id value 1

Listing 10. Slot "1" of Cisco 7603.

Suggest allowed card categories. Supporting the user by providing suggestions about the allowed card categories that can be plugged into a slot in a given configuration instance is the guidance service required by the use case UC-3. The service is realised via a direct call to the reasoner.

Ontology pre-processing is needed when the ontology already contains a card in the slot being queried. If such axiom exists, it is removed from the ontology.

Subsumption checking is realised via a direct call to the reasoner. Specifically, we look for all descendant classes of the following class: `pd:Card and not (inv(pd:hasCard) value ?slotId)`, where `?slotId` is a placeholder for the slot, which is the parameter of the service. That is, we check for all subclasses of the `pd:Card` for which we can prove that they cannot be inserted into the given slot.

CWA complement computation phase is responsible for transforming the negative answer returned by the subsumption checking (the disallowed `CardCategories`) into positive answer required by the user (the allowed `CardCategories`). This complement computation requires the Close World Assumption (CWA).

Result interpretation involves retrieving the `CardCategories` from the PDDSL model that correspond to the set of classes from the ontology, computed in the previous steps. The set of those `CardCategories` is the result returned to the user.

One of the scenarios where the service is used is the model repair process after the detection of an inconsistency. For instance, if we consider the model from Listing 2, the next step after the inconsistency is detected is to ask for the card categories which are allowed in slot "1". For the purpose of this query, the wrong card is removed from the slot and the reasoner is asked for the named subclasses of `pd:Card and not (inv(pd:hasCard) value cisco_7603_slot_1)`. The result of this query is the set of all named classes which are known not to be allowed in the slot. After the complement of this set of classes is computed, the user would get the following set of card categories: `Supervisor_Engine_2`, `Supervisor_Engine_720` as the result of the service.

6 Related Work

Despite the fact that our work has similar motivations and uses the concepts of orthogonal instantiation as the 2D metamodelling architecture defined by Atkinson and Kuehne, there is one fundamental difference in the way this architecture is applied. In [6], the M2 linguistic level does not contain any domain-specific concepts. This implies that the M2 level does not include the definition of the DSL and it simply remains the same for all languages. Thus, in [6] the language engineering has to rely on the ontological metamodelling, and is fully contained within the M1 linguistic layer.

Our approach does not go as far from the typical way of the DSL engineering. We still make use of the linguistic metamodelling to define the concepts of the DSL. We then use ontology technology to give meaning to the ontological instantiation relationship, which would be otherwise treated as an ordinary association by the metamodelling tool. While, in [6] several advantages of the purely ontological metamodelling were presented, such architecture is not easy to apply, making our solution an interesting contribution for the language engineers.

The main reason for that is that our work fits well with the current tools, such as EMF, which enforce purely linguistic metamodelling. Following the typical scenario, the language engineers operate on the M2 level and the language users interact with the M1 level. Thus, our prototype may profit from the support for the linguistic metamodelling implemented in the existing tools. Secondly, the integration with OWL2 is only slightly invasive, and allows for the compatibility with any already existing developments. The language developer can start work with the existing metamodels in order to bridge it with OWL2. The already existing M1 models expressed in the DSL are also preserved without any extra effort. Thus, our approach can be applied to simply enhance already existing tools, languages and models rather than invasively redesign the metamodelling architecture.

The extensive comparison of our work with [6] is out of scope of this paper. However, at least one notable aspect is worth mentioning, i.e. the ability to represent multiple levels of ontological instantiation. The architecture presented in [6] supports the multilevel instantiation. In contrast, our prototype currently allows for only one ontological instantiation. However, the support for more levels of ontological instantiation is also possible to achieve by bridging to the ontological technical space, given the existing work aiming at representing the metamodelling layers in ontologies (OWL2 FA [16]). Also, the semantics of the "instance of" relationship can be explicitly defined. Thus, it is possible to represent the specialized kinds of the "instance of", which could be then used to express different meanings, which would result in a N-dimensional classification architecture.

Finally, since the support for ontological instantiation is just one example of usage of the ontology technology in the language engineering, it can be regarded as an application of a more general mechanism, i.e. the formalization of the semantics of the DSL constructs.

7   Summary

Even though ontologies have been used for a long time in computer science, integration with the domain specific languages has only been investigated to a limited extent in the field of data modelling. The problems described and appearing in everyday tasks are of an abstract nature and cannot be easily solved using the existing tools and approaches. Introducing the integrated models which combine structural and semantic information is surely a great advantage, and will lead to the improvement of the existing systems, making them more user-friendly. The presented usage scenario serves as a proof of concept for the ontology-enriched modelling. The initial results have already proved its usefulness.

Our prototypical approach enables the formalisation of the semantics of modelling language, using a combination of the metamodel integration and model transformation technologies. The resulting ontology can then be reused for various purposes (validation of the type model, validation of the instances against their types and providing suggestions given the current state of the model). The implementation of such use cases would require hard-coding the semantics within the tools that interact with the language.

With our approach, we show that it is possible to combine metamodelling and ontologies to handle such cases. The metamodelling (linguistic) aspect defines the domain specific language for both the type layer and the instance layer. The language definition includes the metaproperty which represents instantiation (hasType). Then, having defined the bridge to the ontology, the semantics of such relationship can be defined as the class assertion. Thus, the validation of the instances against types can be performed, as well as many other use cases.

Furthermore, the prototype allows for combining the DSL constructs with the OWL2 axioms in the PDDSL models. We have evaluated this approach in a scenario of representation of the compatibility constraints between the categories of the cards. Without the integrated approach, such constraints would require the extension of the PDDSL language definition. Our observation is that the language concepts should be able to express the typical cases found in the domain, while the rare cases can be expressed by the OWL2 axioms. This way the DSL is kept simple while not limited in the expressiveness.

We believe that our work, base on the specific case study, can be generalized in order to support broader range of cases where the 2D modelling is useful. Thus, our observation is that our approach can significantly contribute to the DSL engineering.

References

1. PlanetMDE: A Web Portal for The Model Driven Engineering Community (2010), http://planetmde.org
2. Wende, C., Bartho Andreas Ebert, J., Jekjantuk, N., Gerd, G., Lemke, J., Miska, K., Rahmani, T., Sabina, P., Schwarz, H., Walter, T., Zhao, Y., Zivkovic, S.: D2.5 - ontology services for model-driven software development. MOST Project Deliverable (November 2009)
3. Forum, D.: DSM Forum web page (2010), http://www.dsmforum.org/
4. Fleck, J.: Overview of the Structure of the NGOSS Architecture (2003)
5. Atkinson, C., Kühne, T.: Model-driven development: A metamodeling foundation. IEEE Software (September/October 2003)
6. Atkinson, C., Gutheil, M., Kennel, B.: A flexible infrastructure for multilevel language engineering. IEEE Transactions on Software Engineering 99(RapidPosts), 742–755 (2009)
7. EMF: Eclipse Modelling Framework Web Page (2010) http://www.eclipse.org/emf
8. EMFText: EMFText Web Page (2010) http://www.emftext.org
9. Parreiras, F.S., Staab, S., Winter, A.: On marrying ontological and metamodeling technical spaces. In: ESEC-FSE '07: Proceedings of the the 6th Joint Meeting of the

European Software Engineering Conference and the ACM SIGSOFT Symposium on the Foundations of Software Engineering, pp. 439–448. ACM, New York (2007)

10. Silva Parreiras, F., Staab, S., Winter, A.: TwoUse: Integrating UML Models and OWL Ontologies. Technical Report 16/2007, Universität Koblenz-Landau, Fachbereich Informatik (2007)

11. Horridge, M., Patel-Schneider, P.F.: OWL 2 Web Ontology Language Manchester Syntax. Technical report (2009),
    http://www.w3.org/TR/owl2-manchester-syntax/

12. OMG: MOF QVT Final Adopted Specification (2005),
    http://www.omg.org/docs/ptc/05-11-01.pdf

13. Damásio, C.V., Analyti, A., Antoniou, G., Wagner, G.: Supporting Open and Closed World Reasoning on the Web. In: Alferes, J.J., Bailey, J., May, W., Schwertel, U. (eds.) PPSWR 2006. LNCS, vol. 4187, pp. 149–163. Springer, Heidelberg (2006)

14. Yuan, R., Pan, J.Z., Zhao, Y.: Soundness preserving approximation for tbox reasoning (2010)

15. Haase, P., Qi, G.: An analysis of approaches to resolving inconsistencies in dl-based ontologies (2007), http://kmi.open.ac.uk/events/iwod/papers/paper-13.pdf

16. Pan, J.Z., Horrocks, I., Schreiber, G.: OWL FA: A Metamodeling Extension of OWL DL. In: Proc. of the First International OWL Experience and Directions Workshop, OWLED 2005 (2005)

Bridging Query Languages in Semantic and Graph Technologies*

Hannes Schwarz and Jürgen Ebert

Institute for Software Technology, University of Koblenz-Landau
{hschwarz,ebert}@uni-koblenz.de

Abstract. Software systems and software development itself make use of various technological spaces, e.g., relational databases, XML technology, ontologies, and programming languages. The usage of several technological spaces in a single system or development process requires a proper *bridging* between them. "Bridging" can be achieved by transforming concepts of one space into another, by using an adapter to make concepts of one space usable in another, or even by integrating both spaces into a single one, for instance.

This paper presents a transformation-based bridge between the query languages SPARQL and GReQL, with SPARQL originating from the semantic technological space and GReQL representing the model-based space. Since transformation of queries requires the prior mapping of the queried knowledge base, the approach also involves transforming the underlying data in the form of RDF documents and TGraphs, respectively. The benefits of the bridge are shown by applying it for traceability.

1 Introduction

Software systems often make use of different technologies to represent data, e.g., object-oriented programming and relational databases. Similarly, software development itself employs various technologies and languages to create a large body of artifacts. As further detailed below, this paper deals with an approach to *"bridge"* query languages of two of these technologies, i.e., to make it possible to use them cooperatively in a single system or development project.

Bridging technological spaces. Kurtev et al. coined the notion of *technological spaces* to distinguish between different technologies or families of languages [1]. Sample technological spaces are programming languages, XML, semantic technologies including RDF [2] and OWL 2 [3], and relational databases. There is no common agreement or standard on the categorization or naming of technological spaces. Instead, the term serves as theoretical, rather "informal" concept to classify such a "working context with a set of associated concepts, body of knowledge, tools, required skills, and possibilities" [1].

---

* This work has been funded by the European Commission within the 7th Framework Programme project MOST no. ICT-2008-216691, http://most-project.eu.

U. Aßmann, A. Bartho, and C. Wende (Eds.): Reasoning Web 2010, LNCS 6325, pp. 119–160, 2010.

The combined usage of different technological spaces often requires to bridge them. In model-driven development, for example, such bridges take the form of transformations of concepts from one modeling approach to another one. Besides using transformations, it is possible to bridge technological spaces by adapters making concepts of one space usable in another, or by the integration of different technological spaces into a single one, for instance.

Motivation and goal. A currently beginning trend in research is to bridge the *semantic* and the *model-based* technological spaces in order to be able to employ the services of ontologies and reasoning in software development, which usually relies on modeling approaches such as MOF [4] or Ecore [5]. For example, the MOST project[1] has the goal to establish various such bridges to facilitate *ontology-driven software development*. Embedded in the context of MOST, this paper presents a transformation-based bridge between two query languages of the semantic and model-based technological spaces: the currently recommended standard RDF query language *SPARQL 1.0* [6] and the graph query language *GReQL* [7], respectively. When transforming artifacts and models from one space into the other to make use of proprietary services, the approach relieves developers from the need to manually rewrite already formulated queries.

Since the execution of a transformed query requires the prior transformation of the queried data structures, this work also includes a mapping between the relevant approaches: *RDF* and *TGraphs*. Compared to other approaches of the model-based space such as MOF or Ecore, the TGraph approach including GReQL offers decisive advantages for various applications, e.g., traceability [8]. Thus, it is chosen as representative for the model-based space in this work.

The usefulness of the bridging approach is exemplified by an application in the field of traceability: it is shown how to employ GReQL to query traceability information represented in an ontology. Other, comparable applications relying on expressive graph queries are similarly supported.

Assumption for the bridging approach. It is important to understand that the introduced approach for transforming RDF data to TGraphs and vice versa only results in a conversion of the data structures, i.e., it is purely syntactical without preserving the underlying semantics. Most importantly, RDF adopts the *open world assumption*, i.e., information which is not explicitly specified is not necessarily non-existent. Furthermore, explicitly specified information may *entail* additional information which can be inferred and materialized using reasoning approaches. In contrast, the TGraph approach is based on the *closed world assumption*, i.e., all information which is not explicitly contained in a graph is assumed to not exist and cannot be queried by GReQL. However, SPARQL (1.0) per se – without prior reasoning – behaves similarly to GReQL: it only retrieves information explicitly captured by an RDF construct. Consequently, the differences in the semantics of RDF and TGraphs do not bear any relevance to the pursued bridging approach between the query languages.

---

[1] http://www.most-project.eu

Structure of this paper. This paper is structured as follows: in Section 2, RDF and SPARQL are introduced. The TGraph approach with GReQL as a representative for the model-based technological space is described in Section 3, including a discussion of the approach's advantages over MOF with its query language OCL [9]. While Section 4 explains the mapping between RDF and TGraphs, Section 5 shows how to transform SPARQL queries to GReQL queries and vice versa. In Section 6, applications of the approach are discussed, including the aforementioned application in the context of traceability. Subsequently, Section 7 gives a short overview of related work. Finally, Section 8 concludes this paper and offers an outlook on future work.

2 Semantic Technology: RDF and SPARQL

RDF, short for *Resource Description Framework*, is a World Wide Web Consortium (W3C) standard for describing and structuring information on the WWW. SPARQL is a recursive acronym for *SPARQL Protocol and RDF Query Language* and refers to the standard query language for RDF data. Since the syntax of OWL (*Web Ontology Language*), another W3C standard, can be serialized to RDF, SPARQL is widely used as query language for ontologies. Therefore, RDF and SPARQL have been chosen to represent the semantic technological space.

The following Sections 2.1 and 2.2 introduce RDF and SPARQL in a level of detail sufficient to comprehend the bridging approach in this paper.

2.1 RDF – Resource Description Framework

RDF describes information about and relations between *resources*. A resource is "something" which is identifiable on the World Wide Web, e.g., a Web page, an item in an online shop, or an entry in an address book [10].

RDF basics. An RDF document is basically a set of so-called *triples* (or, synonymously, *statements*) consisting of a *subject*, a *predicate*, and an *object*. A set of triples can be represented as an *RDF graph* with the subjects and objects constituting the *nodes* and the predicates constituting the *arcs*. Arcs are directed: they start at the subject node and end at the object node.

Nodes can be either resources, globally identifiable by a *URI reference*, they can be *blank* without any means to globally identify them, or they can be a *literal*, i.e., a data value. However, literals cannot act as subjects in RDF triples [11]. Predicates are occurrences of resources and are therefore identifiable by a URI reference. Within an RDF document, *blank node identifiers* can be used to locally refer to specific blank nodes. In the example RDF document in Figure 1, _:ad1 is an identifier for a blank node aggregating the address data, represented as *plain literals*. URI references are abbreviated by *prefixes*.

Plain literals are essentially simple strings, optionally accompanied by a tag denoting the language of the text. The set of all plain literals is subsumed by

```
@prefix addressBook: <http://de.uni_koblenz.rdf/addrBook#> .
@prefix personal: <http://de.uni_koblenz.rdf/personalAddrBook#> .

personal:entry1 addressBook:firstName "Lisa" .
personal:entry1 addressBook:address _:ad1 .
_:ad1 addressBook:street "Unter_den_Linden" .
_:ad1 addressBook:postalCode "10117" .
_:ad1 addressBook:city "Berlin" .
_:ad1 addressBook:country "Germany" .
```

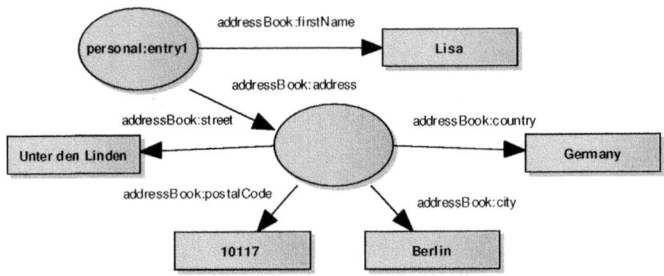

Fig. 1. RDF document example with corresponding RDF graph

the predefined datatype rdf:PlainLiteral[2]. Another kind of literals, *typed literals*, is also represented as string, but is followed by the URI of a datatype the string is to be interpreted as. For typed literals, RDF itself only defines a single datatype: rdf:XMLLiteral, representing XML content. However, since *XML schema* datatypes [12] are commonly used in RDF, it is legitimate to consider them as inherent part of RDF. An overview can be found in Figure 2.

To specify instance-of relationships, RDF predefines the rdf:type resource. Triples with rdf:type as predicate define that the subject is an instance if the resource acting as object. Instances of the predefined resource rdf:Property are the resources the predicates are occurrences of. Although usually, *property* and *predicate* are used synonymously [2], their distinction is important for the purposes of this paper, so that to avoid confusion, *property* will not be used.

Other RDF resources are rdf:Bag, rdf:Seq, rdf:Alt, and rdf:List. The first three resources specify different types of *containers*: rdf:Bags can contain specific resources and literals multiple times. rdf:Seqs additionally impose an ordering on their members. rdf:Alts contain alternative resources or literals. rdf:Lists are ordered, so-called *collections* of resources and/or literals which may occur more than once in one rdf:List. Neither containers nor collections are homogeneous, i.e., their members do not have to be of the same type.

RDF Schema. The RDF vocabulary is extended by *RDF Schema* [13], which defines resources such as rdfs:Resource[3], rdfs:Literal, rdfs:Class, rdfs:subClassOf,

---

[2] The prefix rdf abbreviates the URI reference
   <http://www.w3.org/1999/02/22-rdf-syntax-ns#> .
[3] The prefix rdfs abbreviates the URI reference
   <http://www.w3.org/2000/01/rdf-schema#>.

RDF datatype	Comment
rdf:PlainLiteral	represents plain literals
rdf:XMLLiteral	represents XML content
xsd:anyURI	represents a URI reference
xsd:base64Binary	represents Base64-encoded binary data
xsd:boolean	represents a boolean value
xsd:date	represents a calendar date
xsd:dateTime	represents a point in time
xsd:decimal	represents a decimal number of arbitrary precision
xsd:double	represents a 64-bit signed floating point number
xsd:float	represents a 32-bit signed floating point number
xsd:gDay	represents a day recurring every month in every year
xsd:gMonth	represents a month recurring every year
xsd:gMonthDay	represents a day in specific month recurring every year
xsd:gYear	represents a calendar year
xsd:gYearMonth	represents a month in a specific year
xsd:hexBinary	represents hex-encoded binary data
xsd:int	represents a 32-bit signed integer number
xsd:long	represents a 64-bit signed integer number
xsd:string	represents a character string
xsd:time	represents a point in time recurring every day

Fig. 2. RDF datatypes [12], excluding most subtypes of xsd:decimal and xsd:string

rdfs:subPropertyOf, rdfs:domain, and rdfs:range. Resources and literals are instances of rdfs:Resource and rdfs:Literal, respectively, with both of the latter being instances of rdfs:Class. Specialization relationships between rdfs:Classes and rdf:Propertys are described by the rdfs:subClassOf and rdfs:subPropertyOf properties, respectively. All classes, user-defined ones as well as rdfs:Class, rdfs:Literal, and rdfs:Property are instances of rdfs:Class and subclasses of rdfs:Resource.

If the subjects of triples with a rdfs:domain predicate occur as predicates in other triples, the subjects of the latter are instances of the class which is the object of the respective rdfs:domain triple. Similarly, rdfs:range triples imply that if their subjects are used as predicates in other triples, the objects of the latter are instances of the class which is the object of the respective rdfs:range triple.

Entailment. Entailment denotes the inference of new triples from existing triples to make previously implicit knowledge explicit. An example is the classification of a predicate being the occurrence of an rdf:Property instance. For instance, the triple "A p B" entails "p rdf:type rdf:Property". The specific set of rules which is applied to infer triples is identified as *entailment regime*.

The elementary entailment regime is called *simple entailment*. Basically, simple entailment only comprises the replacement of blank node identifiers by a resource or literal. Other, more sophisticated regimes such as *RDF entailment*, *RDFS entailment* and *D-Entailment* consider the semantics of RDF concepts, RDF Schema concepts, and datatypes, respectively [11].

2.2 SPARQL – SPARQL Protocol and RDF Query Language

The current W3C recommendation for the RDF query language SPARQL [6] corresponds to version 1.0 and constitutes the basis for the following language description. SPARQL 1.0 is designed for simple entailment only and does not support inferencing going further than the replacement of variables. Thus, SPARQL exhibits closed world semantics, i.e., it is only able to retrieve explicitly specified RDF data.

Central language features of SPARQL are the usage of *triple patterns* resembling RDF triples to specify the desired result, the ability to construct new RDF graphs, and the possibility to query more than one RDF graph in a single query. SPARQL queries can consist of up to five query parts:

```
prologue
query form
dataset
where clause
solution modifiers
```

Depending on the query form, some parts may be optional. Details on this are given in the subsequent paragraphs dealing with the different query parts.

Prologue. The prologue allows to specify a *base IRI*[4] and an arbitrary number of prefix-IRI assignments. A base IRI allows to use relative IRIs within a SPARQL query: when using only the local part of an IRI, the base IRI is prepended. Prefix specifications allow to use shorter prefixes instead of the whole IRI. Specifying a prologue is always optional. An example is displayed below.

```
BASE <http://de.uni_koblenz.rdf/addressBook#>
PREFIX rdf: <http://www.w3.org/1999/02/22-rdf-syntax-ns#>
```

With this prologue, the IRI <http://de.uni_koblenz.rdf/addrBook# entry1> can be referred to by <entry1>. Further, rdf:type can be used for <http://www.w3.org/1999/02/22-rdf-syntax-ns#type>.

Query form. SPARQL offers four different kinds of query forms: SELECT, ASK, CONSTRUCT, and DESCRIBE. To understand the purpose of each query form, it is necessary to introduce the notion of *solution mappings* which bind variables to *RDF terms*, i.e., IRIs, blank nodes, or literals. More precisely, a solution mapping is a partial function $\mu : V \nrightarrow T$. For a given query, V corresponds to the set of all variables used in the query and T represents the entirety of RDF terms in the queried RDF graphs.

SELECT queries return a bag of *solution mappings* Ω. It is possible to *project* V to a set $V_p \subseteq V$ to be included in the query's result. The result bag is represented as a table with all variables in V_p as column titles and the rows

---

[4] IRI is short for *Internationalized Resource Identifier*, an internationalized version of URIs. For the purposes of this document, the difference is of no relevance.

representing solution mappings. Each table cell contains the value $\mu(v)$, with $\mu \in \Omega$ being the solution mapping represented by the row of the table cell and $v \in V_p$ being the variable of the respective column. If $\mu(v)$ is not defined for a solution mapping, the according cell is left empty. The RDF terms each variable is bound to are determined by the triple patterns in the where clause (see below). Identical solution mappings can be eliminated by using the DISTINCT or REDUCED keywords after SELECT. While DISTINCT *ensures* the removal of duplicate solutions, the usage of REDUCED only *allows* to remove them.

ASK queries check whether there exists any solution mapping for the structures given in the where clause and returns the respective boolean value.

CONSTRUCT queries serve to build and return new RDF graphs using a template in the form of triple patterns. The values of the solution mappings fulfilling the triple patterns given in the where clause form the triples of the new graph.

DESCRIBE queries return an RDF graph describing resources. In contrast to the CONSTRUCT query form, the graph's structure is not prescribed by the SPARQL user, but is determined by the provider of the RDF graph.

Specification of the query form is mandatory. Examples for the different query forms are given below when the where clause is discussed.

Dataset. The dataset specifies the RDF graphs on which the SPARQL query is executed. It must be distinguished between the *default graph*, specified by using the FROM keyword, and *named graphs*, specified by FROM NAMED. Triple patterns in the where clause are matched against the default graph unless the GRAPH keyword is used to specify named graphs to be matched against. Specifying more than one IRI for the default graph results in *merging* the individual graphs. A merge basically corresponds to taking the union of several RDF graphs [11].

Specification of the dataset is optional because besides giving the dataset in the query, the dataset can also be specified by the SPARQL protocol facilitating the communication between a query client and a query processor. A dataset specification in the protocol overrides any datasets in the query.

Where clause. The where clause consists of triple patterns specifying the eligible structures to be included in the result of the query. Its usage is mandatory except for DESCRIBE queries. Several example queries in this section show the various SPARQL features usable in the where clause. The following two RDF documents constitute the knowledge base on which the queries operate:

```
# URI: http://de.uni_koblenz.rdf/personalAddrBook
@prefix rdf: <http://www.w3.org/1999/02/22-rdf-syntax-ns#>
@prefix xsd: <http://www.w3.org/2001/XMLSchema#>
@prefix addressBook: <http://de.uni_koblenz.rdf/addrBook#> .
@prefix personal: <http://de.uni_koblenz.rdf/personalAddrBook#> .

personal:entry1 rdf:type addressBook:Person .
personal:entry1 addressBook:name "Lisa" .
personal:entry1 addressBook:age "28"^^xsd:int .
personal:entry1 addressBook:address _:ad1 .
_:ad1 addressBook:street "Unter_den_Linden" .
_:ad1 addressBook:postcode "10117" .
_:ad1 addressBook:city "Berlin" .
_:ad1 addressBook:country "Germany" .
```

```
personal:entry2 rdf:type addressBook:Person .
personal:entry2 addressBook:name "Hugo" .
personal:entry2 addressBook:age "26"^^xsd:int .
personal:entry2 addressBook:address _:ad2 .
_:ad2 addressBook:street "Avenue_des_Champs-Elysees" .
_:ad2 addressBook:postcode "75008" .
_:ad2 addressBook:city "Paris" .
```

```
# URI: http://de.uni_koblenz.rdf/businessAddrBook
@prefix rdf: <http://www.w3.org/1999/02/22-rdf-syntax-ns#>
@prefix addressBook: <http://de.uni_koblenz.rdf/addrBook#> .
@prefix business: <http://de.uni_koblenz.rdf/businessAddrBook#> .

business:entry1 rdf:type addressBook:Organization .
business:entry1 addressBook:name "MIB_SE" .
business:entry1 addressBook:address _:ad1 .
_:ad1 addressBook:street "Via_Appia_Antica" .
_:ad1 addressBook:zip "00179" .
_:ad1 addressBook:city "Rome" .
_:ad1 addressBook:country "Italy" .
```

The simplest queries ask for the existence of specific triples and return the appropriate boolean value. More sophisticated queries are allowed by using variables in triple patterns, identifiable by a preceding question mark. The query below asks whether the queried graphs contain a single RDF term acting as subject for some triple with the predicate addressBook:city and the plain literal "Paris" as object, *and* some triple with the predicate addressBook:city and the plain literal "Rome" as object. Obviously, the result is no.

```
PREFIX rdf: <http://www.w3.org/1999/02/22-rdf-syntax-ns#>
PREFIX addressBook: <http://de.uni_koblenz.rdf/addrBook#>

ASK
FROM <http://de.uni_koblenz.rdf/personalAddrBook>
WHERE {
    ?address addressBook:city "Paris" .
    ?address addressBook:city "Rome" .
}
```

In the above example, it is asked whether there is any solution mapping fulfilling *all* the triple patterns in the where clause. Regarding the two patterns individually, the only solution mapping μ_1 for the first pattern is $\mu_1(?address) = \_ : \mathtt{ad1}$, while the only mapping for the second pattern is $\mu_2(?address) = \_ : \mathtt{ad2}$. Thus, $\Omega_1 = \{\mu_1\}$ and $\Omega_2 = \{\mu_2\}$. The bag of solution mappings for the whole query is the *join* ($\bowtie$) of Ω_1 and Ω_2. The join is defined as follows:

$$\Omega_1 \bowtie \Omega_2 = \{\mu_1 \cup \mu_2 \mid \mu_1 \in \Omega_1, \mu_2 \in \Omega_2, \forall\, v \in dom(\mu_1) \cap dom(\mu_2) : $$
$$\mu_1(v) = \mu_2(v)\}$$

For the example, the result of computing the join of Ω_1 and Ω_2 is the empty set.

With the SELECT query form, solution mappings are returned in tabular form. If the queried graphs match the triple patterns in the where clause, i.e., if the graphs contain the triples resulting from replacing variables by RDF terms, the respective variable bindings form a table row. If not specified otherwise, all triple patterns have to be matched. The following query returns a table whose rows correspond to the *name-street* combinations which can be extracted from the address book. The resulting table is displayed below the query.

```
PREFIX rdf: <http://www.w3.org/1999/02/22-rdf-syntax-ns#>
PREFIX addressBook: <http://de.uni_koblenz.rdf/addrBook#>

SELECT ?name ?street
FROM <http://de.uni_koblenz.rdf/personalAddrBook>
WHERE {
    ?entry addressBook:name ?name
    ?entry addressBook:address ?address .
    ?address addressBook:street ?street .
}
```

name	street
"Lisa"	"Unter den Linden"
"Hugo"	"Avenue des Champs-Elysees"

Variables are also used in CONSTRUCT queries to describe the RDF terms used to build a graph. In the example below, the UNION keyword is employed to denote that the two groups of triple patterns enclosed by braces – so-called *group graph patterns* – are alternatives. Thus, solution mappings for either of the the group patterns are returned. Since there is no address blank node participating in both zip and postcode triples, the resulting graph would be empty without UNION.

```
PREFIX rdf: <http://www.w3.org/1999/02/22-rdf-syntax-ns#>
PREFIX addressBook: <http://de.uni_koblenz.rdf/addrBook#>
PREFIX zipLookup: <http://de.uni_koblenz.rdf/zipLookup#>

CONSTRUCT { ?zip zipLookup:city ?city }
FROM <http://de.uni_koblenz.rdf/businessAddrBook>
FROM <http://de.uni_koblenz.rdf/personalAddrBook>
WHERE {
    {
        ?address addressBook:zip ?zip .
        ?address addressBook:city ?city .
    } UNION {
        ?address addressBook:postcode ?zip .
        ?address addressBook:city ?city .
    }
}
```

```
@prefix zipLookup:    <http://de.uni_koblenz.rdf/zipLookup#> .

"10117" zipLookup:city "Berlin"
"75008" zipLookup:city "Paris"
"00179" zipLookup:city "Rome"
```

The union ($\cup$) of two bags of solution mappings Ω_1 and Ω_2 is defined as follows:

$$\Omega_1 \cup \Omega_2 = \{\mu \mid \mu \in \Omega_1 \vee \mu_2 \in \Omega_2\}$$

The definition implies that if both alternatives denoted by UNION match the same triples in the queried graph, the respective solutions are returned twice.

The OPTIONAL keyword is used to specify triple patterns which are not mandatory to be matched. In the example below, bindings for the variable ?name are included in the result table even if there is no suitable binding for ?age. The example also illustrates the usage of FILTERs for restricting solution mappings to those causing the associated expression to evaluate to true. SPARQL provides a variety of operators and functions, partially imported from XQuery and XPath [14], to be used in FILTER expressions. Examples are arithmetic and logical operators or regular expression matching. In addition to predefined operators and functions, SPARQL allows to refer to user-defined functions.

```
PREFIX rdf: <http://www.w3.org/1999/02/22-rdf-syntax-ns#>
PREFIX addressBook: <http://de.uni_koblenz.rdf/addrBook#>

SELECT ?name ?age
FROM <http://de.uni_koblenz.rdf/personalAddrBook>
FROM <http://de.uni_koblenz.rdf/businessAddrBook>
WHERE {
    ?entry addressBook:name ?name .
    OPTIONAL {
        ?entry addressBook:age ?age .
        FILTER (?age > 27)
    }
}
```

name	age
"Lisa"	"28"^^xsd:int
"Hugo"	
"MIB SE"	

Basically, the bag of solution mappings fulfilling the patterns in the where clause is computed by first taking the bag of solution mappings resulting from the join of the mappings fulfilling the non-optional patterns (Ω_1) with the mappings fulfilling the optional patterns (Ω_2), and, second, unifying this bag with the bag of all mappings μ_3 from Ω_1 so that for all mappings μ_4 in Ω_2 which have variables in common with μ_3, there must be at least one variable v with $\mu_3(v) \neq \mu_4(v)$. This operation is also described as *left (outer) join*:

$$\Omega_1 \text{ LEFTJOIN } \Omega_2 = (\Omega_1 \bowtie \Omega_2) \cup$$
$$\{\mu_3 \in \Omega_1 \mid \forall \mu_4 \in \Omega_2 : \exists v \in dom(\mu_3) \cap dom(\mu_4) :$$
$$\mu_3(v) \neq \mu_4(v)\}$$

The OPTIONAL keyword is left-associative, i.e., a sequence of optional group graph patterns is evaluated in the order of their occurrence. Other orders can be achieved by using nested group graph patterns. In the following example, without the braces surrounding the two innermost triple patterns, there would be no binding for ?city in the second row of the result table.

```
PREFIX rdf: <http://www.w3.org/1999/02/22-rdf-syntax-ns#>
PREFIX addressBook: <http://de.uni_koblenz.rdf/addrBook#>

SELECT ?name ?country ?city
FROM <http://de.uni_koblenz.rdf/personalAddrBook>
FROM <http://de.uni_koblenz.rdf/businessAddrBook>
```

```
WHERE {
    ?entry addressBook:name ?name .
    ?entry addressBook:address ?address .
    OPTIONAL {
        {
            ?address addressBook:country ?country
            OPTIONAL { ?address addressBook:city ?city }
        }
    }
}
```

name	country	city
"Lisa"	"Germany"	"Berlin"
"Hugo"		"Paris"
"MIB SE"	"Italy"	"Rome"

Solution modifiers. Solution modifiers are employed to sort or to restrict the number of computed solution mappings. Sorting in ascending or descending order can be achieved by the ORDER BY keyword together with ASC or DESC, respectively, followed by one or more variables as sorting criteria. The number of generated solution mappings is controlled with the OFFSET and LIMIT keywords followed by a number. OFFSET x specifies that the first x solutions found are not returned. LIMIT y ensures that no more than y solutions are returned.

The usage of solution modifiers is optional except for ASK queries, where they must not be specified.

3   Model-Based Technology: The TGraph Approach

The approach to graph-based modeling using TGraphs has been used for various applications, including the development of meta-case tools [15], reengineering [16], and the storage of traceability information [17]. While TGraphs and the query language GReQL are in most aspects comparable to the well-established OMG standards MOF and OCL, some essential features not supported by the latter render the TGraph approach superior for the purposes of this work.

Section 3.1 discusses TGraphs themselves and the *grUML (graph UML)* language as a means to model TGraph schemas. Section 3.2 describes the complementing query language GReQL, short for *Graph Repository Query Language*. In Section 3.3, the benefits of the TGraph approach including GReQL are illustrated by briefly comparing it to EMOF (Essential MOF) and OCL.

3.1   TGraphs and grUML

The TGraph approach adheres to a four-layered meta-hierarchy comparable to MOF. TGraphs themselves reside on the M1 level (with run-time instances on M0). Their meta-models, called *schemas*, situated on the M2 level and modeled in grUML, again conform to a defined *meta-schema* on the M3 level.

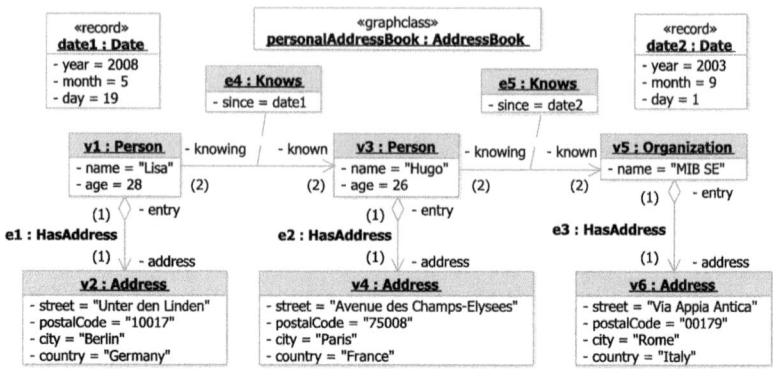

Fig. 3. TGraph example

TGraphs. TGraphs are a very general kind of graphs. Their main four properties are that they are:

- *typed*, i.e., vertices, edges, and the graph itself have a type,
- *attributed*, i.e., vertices, edges, and the graph can carry attribute–value pairs,
- *directed*, i.e., edges have a start and an end vertex,
- *ordered*, i.e., vertices and edges are globally ordered within the graph. Further, the incident edges of a vertex are ordered.

In addition, edges are first class citizens, i.e., they are treated equally with vertices. This includes that they can be explicitly referred to store them in variables.

Figure 3 shows a TGraph counterpart of the RDF graph in Section 2.1, extended by a *knows* relationship to better exemplify the properties of TGraphs. The type of the graph itself is AddressBook. Its vertices are of different types: Person, Organization, and Address. Similarly, edges are either instances of HasAddress or Knows. Depending on their type, graph elements possess different attributes. The since attributes of Knows edges are of composite types, so-called *records*.

Edge directions are displayed as arrowheads. The global orders of vertices and edges are represented, for illustrative purposes, by their identifiers v1 to v6 and e1 to e5. The order of a vertex' incident edges is annotated by the numbers in parentheses. While being insignificant in the address book example, incidence orders may play an important role in other applications. For example, the sequence of keywords and expressions in a code statement can be represented by the incidence order of the edges connecting the statement vertex to its parts.

grUML. Figure 4 shows a schema for the sample graph in Figure 3. The grUML language for modeling TGraph schemas is essentially a subset of UML class diagrams. It consists of those constructs which possess a graph-like semantics: while vertex types and their attributes are represented by classes with attributes, edge types are modeled as associations. Edge types may exhibit aggregation or composition semantics. Further, if an edge type features attributes, association classes are used. Since edges cannot be connected by edges, association classes

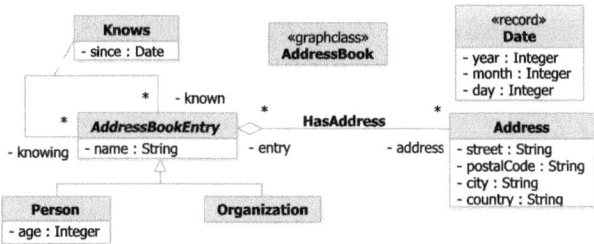

Fig. 4. TGraph schema example in grUML

may not be connected by associations. The type of the graphs conforming to a schema is prescribed by a separate class with the graphclass stereotype.

Generalization relationships are allowed between vertex classes and between edge classes. The semantics of generalization is similar to UML, i.e., specializations inherit their supertypes' attributes and edges are able to connect instances of subtypes of their edge class's incident vertex classes. The direction of edges is determined by the reading direction of the associations representing edge classes (not depicted in Figure 4). Finally, multiplicities denote degree restrictions, i.e., they limit the number of neighboring vertices a vertex may have.

Figure 5 shows a somewhat simplified version of the grUML meta-schema. Apart from VertexClass, EdgeClass, and GraphClass, which are generalized by AttributedElementClass, it defines further concepts: besides the Schema itself, identifiable by schemaPrefix and name and indirectly containing other elements via a default Package, additional Packages can be used for structuring. The edge classes denoted as associations in grUML are actually the concrete syntax for an instance of EdgeClass together with two Incidences as starting and ending points. The possible domains of Attributes belonging to AttributedElementClasses, i.e., specializations of the abstract meta-class Domain are listed in Figure 6.

A TGraph conforms to a schema if the graph, vertex, and edge types together with the attribute assignments obey the specifications of the schema. Further, edges may only be incident to instances of those vertex classes prescribed by the schema and vertex degrees have to respect the given multiplicities.

3.2   GReQL – Graph Repository Query Language

GReQL is an expression language specifically designed for querying TGraphs which also allows for the retrieval of information on TGraph schemas. The term *expression language* refers to the characteristic that every language element is an expression and, obeying certain rules, can be used in other expressions. Furthermore, various functions covering aspects such as logics, arithmetics, collections (e.g., sets or lists), as well as path and graph analysis are available. GReQL's function library contains about 100 functions and can be extended by users.

GReQL can be roughly compared with query languages such as *RPL* [18], *PQL* [19], or even *XPath* [20] which works on XML trees. However, the latter two languages do not support a concept similar to GReQL's regular path expressions

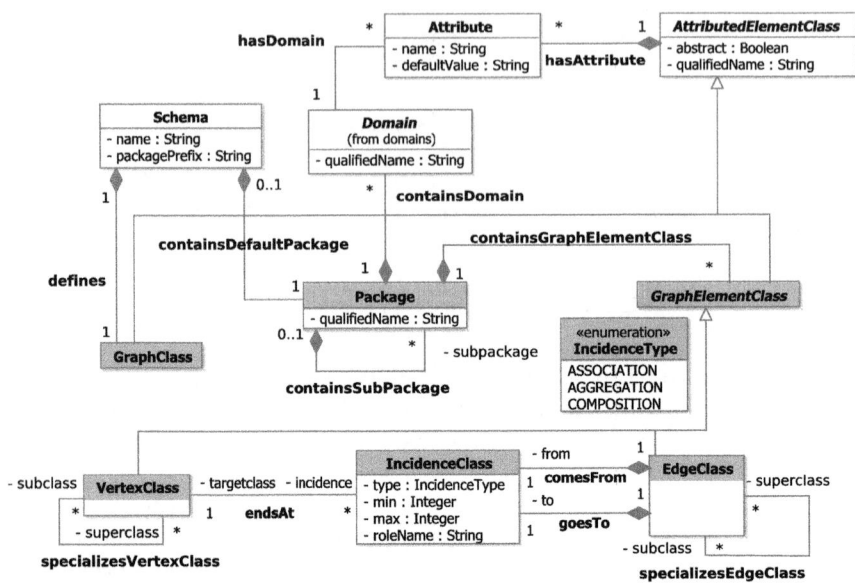

Fig. 5. grUML meta-schema (simplified)

grUML domain	Comment
Boolean	represents a boolean value
Integer	represents a signed 32-bit integer number
Long	represents a signed 64-bit integer number
Double	represents a signed 64-bit floating point number
String	represents a string of any length
List<Domain>	represents a ordered sequence of values of the domain Base-Domain (with the possibility of multiple occurrences)
Set<Domain>	represents an (unordered) set of values of the domain Base-Domain (without multiple occurrences)
Map<Domain,Domain>	represents a mapping of values of the first Domain (the *key domain*) to values of the second Domain (the *value domain*)
Enum	represents a user-defined domain enumerating possible attribute values
Record	represents a user-defined domain composed of any number of *identifier–domain* pairs

Fig. 6. grUML domains

as explained below and thus are less expressive with respect to navigating the queried structures. Further work on comparing GReQL with other (graph) query languages can be found in [7].

Due to the multitude of different GReQL expressions and functions, the following descriptions are restricted to features relevant with respect to the bridge between SPARQL and GReQL. These features include regular path expressions (RPEs), from-with-report expressions, and quantified expressions.

Regular path expressions. RPEs describe the structure of paths in a graph. They are not used as stand-alone expressions, but are embedded in other expressions, most importantly in *path existence* expressions for checking whether a path with a specific structure exists between two vertices, or in *forward vertex set* expressions determining the set of vertices which can be reached from a given vertex by such a path. Analogously, *backward vertex set* expressions compute the vertices from which a vertex can be reached. The examples below, referring to the graph in Figure 3, illustrate these applications. For reasons of brevity, identifiers in the examples such as v2 or e1 directly act as variables referring to the respective graph elements. In complete GReQL expressions, this binding of variables to elements has to be taken care of by the user.

The basic elements of RPEs are *simple path descriptions* describing single, optionally directed edges.

```
1  v1 <-> v3
2  v5 <--{HasAddress}
3  -->{Knows @ thisEdge.since.year < 2001} v3
```

The path existence expression in line 1 checks whether v1 and v3 are connected by any single edge (*true*). The forward vertex set expression in line 2 involves a so-called edge restriction and gives back the set of vertices which can be reached from v5 by following a HasAddress edge in backward direction ({v6}). Finally, the backward vertex set expression in line 3 returns all vertices from which v3 is reachable by following a Knows edge in forward direction and whose value of the year component of the since attribute is smaller than 2001 ($\varnothing$).

Using regular expression operations (concatenation, option, grouping, alternation, iteration), simple path expressions build more complex expressions:

```
4  v2 <-- <->{Knows} [-->]
5  (<--{HasAddress} | -->{Knows}) -->{Knows} v5
6  v1 -->{Knows}+ v5
```

Here, line 4 returns the set of vertices reachable from v2 by first following any edge in backward direction, then a Knows edge in any direction, and optionally any edge in forward direction ({v4, v5}). The result of line 5 are those vertices from which v5 is reachable by first following either a HasAddress edge in backward direction or a Knows edge in forward direction and then another Knows edge in forward direction ({v1, v4}). Line 6 checks whether there exists a path of one or more Knows edges in forward direction from v1 to v5 (*true*).

Other important features of RPEs are *intermediate vertex path descriptions* for referring to specific vertices on a path, *edge path descriptions* for referring to specific edges, *start vertex restrictions* for constraining the start vertices of path descriptions, and, analogously, *goal vertex restrictions* which constrain the end vertices of path descriptions:

```
7  | v1 --e4-> v3 <-e5-> v5
8  | v1 -->{Knows}&{Person}*
9  | {Address @ thisVertex.city = "Berlin" or thisVertex.country = "France"}&<-- v3
10 | v1 -->{Knows}&{Organization} -->{Knows} v5
```

Using an intermediate vertex path description and two edge path descriptions, the expression in line 7 checks if there exists a path from v1 via e4 to v3 and further via e5 to v5, with e4 in forward direction and the direction of e5 being insignificant (*true*). Line 8 involves a goal vertex restriction to retrieve the set of vertices reachable via any number of Knows edges ending at an instance of Person ({v3}). The start vertex restriction in line 9 ensures that the backward vertex set only includes instances of Address with city = "Berlin" or country = "France" ({v4}). The expression in line 10 (*true*) illustrates the usage of vertex restrictions in the middle of a path expression. It checks whether there is a path from v1 via a Knows edge, an Organization vertex and another Knows edge to v5 (*false*).

Conditional expressions. Conditional expressions are used to control the evaluation of expressions depending on the result of a previously evaluated boolean expression. The following example illustrates their syntax:

```
hasType(v3, "Address") ? v3.city : v3.name
```

The boolean expression preceding the question mark employs the hasType() function to check if v3 is of type Address. If *true*, the value of its city attribute is returned. If *false*, v3 is of type AddressBookEntry and the value of name is returned. Since in the example, v3 is an instance of Person, "Hugo" is returned.

from-with-report expressions. from-with-report (FWR) expressions bear resemblance to SQL queries. In the from part, variables are bound to domains, e.g., to the set of instances of a vertex class. The with part features a boolean expression which imposes constraints on the eligible variable values. Finally, the report part specifies the structure of the expression's result. Usually, this is a *bag* of tuples. Consider the following example:

```
from p:V{Person}, a:V{Address}
with p -->{HasAddress} a and a.country = "Germany"
report p, a end
```

This query returns a bag of tuples (p, a) with p and a being instances of Person and Address, respectively. p must be connected to a by a HasAddress edge and the country attribute of a must have "Germany" as value. For the graph in Figure 3, the resulting bag would be {(v1, v2)}.

Since edges are treated as first class citizens of TGraphs, the above example could be formulated by using a single variable bound to the set of instances of the HasAddress edge class. The start and end vertices of those edges are accessed by the functions startVertex() and endVertex(), respectively:

```
from h:E{HasAddress}
with endVertex(h).country = "Germany"
report startVertex(h), endVertex(h) end
```

Quantified expressions. Universally, existentially, and uniquely quantified expressions allow to test whether all, some, or exactly one element(s) of a given collection fulfill a specific boolean expression, respectively. The example below illustrates the structure of quantified expressions: following one of the keywords forall, exists, or exists! is one or more declarations. The boolean expression which has to be checked follows after an @-sign. Line 1 ensures that all Address vertices are incident to at least one incoming HasAddress edge (*true*). Line 2 checks whether there is an AddressBookEntry instance which does not have the empty string as value for name (*false*). Finally, line 3 says that there must be exactly one Person vertex with name = "myself" (*false*).

```
1  forall a:V{Address} @ inDegree{HasAddress}(a) > 0
2  exists e:V{AddressBookEntry} @ e.name = ""
3  exists! p:V{Person} @ e.name = "myself"
```

3.3   Comparison to EMOF and OCL

The modeling power of the TGraph approach is somewhat higher than that of EMOF or its adaptation to the Eclipse world – Ecore: while edges are first-class objects in TGraphs, relations between two objects in EMOF are modeled by two properties, one belonging to each object and declared as being the *opposite* of each other. These connections between objects are neither typed nor do they have an identity, entailing that they cannot carry attributes. Furthermore, EMOF lacks the capability to impose orderings on its model elements.

With respect to GReQL and OCL, it can be said that both languages are similar to a great extent, so that expressions in one language can in most cases be translated to the other one. OCL features not present in GReQL are the ability to define *contexts* for expressions and an operation to iterate over collections with the ability to store the result of an expression involving the current element and to reuse this result in the next iteration. The advantages of GReQL are its graph-orientation with the ability to efficiently handle graph structures and the support of RPEs to describe the structure of paths. In particular, the computation of transitive closure is not possible with OCL.

4   Transformation between RDF and TGraphs

Considering the descriptions of RDF and TGraphs in Sections 2 and 3, RDF is more general, but also less structured than the TGraph approach. In particular, RDF does not possess the following structuring features exhibited by TGraphs:

1. TGraph elements are instances of some schema classes, which are in turn instances of a grUML meta-class. RDF is not confined to a given number of "metalevels", i.e., a resource can be an instance of some class, which is in turn instance of another class, etc.
2. Each TGraph attribute has a given type, whereas a comparable rdf:Property instance may occur in multiple triples with literals of different datatypes.

3. TGraph edges are instances of instances of the meta-class EdgeClass and have an own identity. In contrast, predicates can be understood as *occurrences* of instances of rdf:Property without any means to distinguish between occurrences of the same instance.

Bidirectional transformation between RDF and TGraphs requires that RDF graphs are "TGraph-like", i.e., they adhere to the more stringent structuring of TGraphs and do not make use of the additional possibilities of RDF explained above in points 1 and 2. A suitable transformation procedure, so-called *schema-aware mapping*, is given in Section 4.1. Subsequently, Section 4.2 describes a unidirectional RDF-to-TGraph transformation applicable to any RDF graph. Note, that as explained in Section 1, the transformations only consider the syntax of RDF and TGraphs, but do not intend to preserve their semantics, as this is not needed for the purposes of the SPARQL-GReQL bridge. Furthermore, while being possible in general, the mapping of the rdf:Bag, rdf:Seq, rdf:Alt, and rdf:List resources on the RDF side as well as List, Set, or Map attributes on the TGraph side is omitted from the following descriptions due to space restrictions.

4.1   Schema-Aware Mapping between RDF and TGraphs

Schema-aware mapping between TGraph-like RDF graphs and TGraphs works, in principle, by mapping instances of rdfs:Class to vertex classes, instances of rdf:Property to edge classes, resources and blank nodes to vertices, literals to attribute values, and occurrences of rdf:Property instances to edges or attributes. Every element is one-to-one transformed to a concept of the respective other technological space. Consequently, transforming an RDF graph to a TGraph and then applying the inverse transformation yields the original RDF graph[5] and vice versa. However, there is no one-to-one mapping between datatypes and domains, so that their identity is not necessarily maintained.

Restrictions. Due to the differences in the structure of RDF graphs and TGraphs, both have to adhere to certain restrictions to allow for schema-aware mapping. These restrictions, together with an explanation, are given below.

1. Restrictions on RDF graphs:
   (a) The RDF graph must only span two "metalevels", i.e., objects of triples with the rdf:type property do not occur as subjects elsewhere. Exceptions are instances of rdf:Property, which may occur as subjects in rdfs:domain or rdfs:range triples, and triples with rdf:type as predicate marking classes as instances of rdfs:Class and properties as instances of rdf:Property.
   Explanation: TGraphs, like all MOF-like approaches, only feature two meta-levels on which models are freely configurable. The grUML meta-schema on the M3 level is fixed.
   (b) Instances of rdf:Property with a datatype range must not participate as subject in further rdfs:range triples. If the range is not explicitly specified, it must not occur in triples with literal objects of different datatypes.

---

[5] Possibly with some additional rdf:type triples which were only implicit beforehand.

Further, an rdf:Property instance must not occur in triples with object literals as well as in triples with resources or blank nodes as objects.

Explanation: Each attribute has exactly one domain. An instance of rdf:Property is to mapped to either an attribute or an edge class.

(c) Instances of rdf:Property with a datatype range or occurring in triples with literals as objects must not be a subproperty of another property.

Explanation: There is no generalization hierarchy for attributes.

(d) Instances of rdf:Property must not occur as subjects or objects in triples, except for rdfs:subPropertyOf, rdfs:domain, and rdfs:range triples.

Explanation: Edge classes must not be incident to edges or carry attributes.

(e) A resource must be instance of at most one type, i.e., there may be only one triple with a given resource as subject and rdf:type as property.

Explanation: Each TGraph vertex or edge is instance of exactly one class.

(f) If an rdfs:Class instance is specified as domain for a given rdf:Property instance with a datatype range or if one instance of a class occurs in at least one triple with a literal as object and a given rdf:Property instance, then each instance of this class must occur as subject in exactly one triple with that rdf:Property instance as predicate.

Explanation: A graph element has exactly one value for each attribute defined by the class it is an instance of.

2. Restrictions on TGraphs:

(a) Edges must not possess attributes.

Explanation: Occurrences of rdf:Property instances have no identity and cannot be distinguished. Thus, it is not possible for them to act as subjects in triples which relate them to literals representing attribute values.

(b) Attributes must not use the List, Set, Map, and Record domains.

Explanation: RDF does not provide comparable datatypes.

Provided that the above restrictions are met, using schema-aware mapping, any RDF graph can be reversibly transformed to a corresponding TGraph and vice versa.

Mappings. The following table juxtaposes corresponding RDF and TGraph concepts and describes the transformations in both directions. Readers not interested in the details of the transformation can directly proceed to the following example.

RDF concept	**TGraph concept**
URI reference of RDF graph	schema, graph class
⇒ The prefix and names of the schema and the graph class can be derived from the URI reference of the RDF graph itself.	
⇐ The URI reference of the RDF graph is derived from the schema prefix and name.	
	vertex class Vertex, edge class Edge
⇒ A special vertex class, e.g., Vertex, and a special edge class with Vertex as start and end vertex class, e.g., Edge, are created. Vertex has an attribute of the domain String to take the URI references of resources, e.g., uriRef.	
⇐ *These special vertex and edge classes are not transformed to RDF.*	

instance of rdfs:Class	vertex class
⇒ Every rdfs:Class instance is mapped to a vertex class. The URI reference of the rdfs:Class instance is transformed to the vertex class's qualified name.	
⇐ Every vertex class is mapped to a triple as follows: `uri rdf:type rdfs:Class .` uri is derived from the schema prefix and the vertex class's qualified name .	

instance of rdf:Property [with non-datatype range]	edge class
⇒ Every rdf:Property instance whose range is not a datatype is mapped to an edge class. The edge class's qualified name is derived from the URI reference. Its start and end vertex classes correspond to the domain and range of the rdf:Property instance, respectively. If more than one domain and/or range are given, the start and/or end vertex class is Vertex. The multiplicity at both ends is "*". This mapping step is also applied if the range of the rdf:Property instance is not explicitly specified, but it does not occur in triples with literal objects.	

⇐ Every edge class is mapped to triples as follows:
 	
    `uri rdf:type rdf:Property .`
    `uri rdfs:domain startClassUri .`
    `uri rdfs:range endClassUri .`
 	
uri is derived from the schema prefix and the edge class's qualified name. $startClassUri$ and $endClassUri$ are the RDF representations of the start and end vertex classes, respectively.	

instance of rdf:Property [with datatype range]	attribute
⇒ Every rdf:Property instance whose range is a datatype is mapped to an attribute. The URI reference is transformed to the attribute's name. The domain is derived from the RDF datatype according to Figure 7. This mapping is also applied if the range of the rdf:Property instance is not explicitly specified, but if it only occurs in triples with literal objects.	

⇐ Every attribute of a graph element class is mapped to triples as follows:
 	
    `uri rdf:type rdf:Property .`
    `uri rdfs:domain classUri .`
    `uri rdfs:range datatypeUri .`
 	
uri is derived from the schema prefix, the graph element class's qualified name, and the attribute's name. $classUri$ is the RDF representation of the graph element class possessing the attribute. $datatypeUri$ is the URI reference of the datatype suiting the attribute's domain, according to Figure 7.	

occurrence of rdfs:subClassOf or rdfs:subPropertyOf	generalization relationship
⇒ Every occurrence of rdfs:subClassOf or rdfs:subPropertyOf is mapped to a generalization relationship between the vertex classes or edge classes corresponding to the subject and object of the respective triple. Vertex classes or edge classes corresponding to an rdfs:Class instance which is not the subject of a rdfs:subClassOf or rdfs:subPropertyOf triple are specializations of the special graph element classes Vertex and Edge, respectively.	

⇐	Every generalization relationship between two vertex classes is mapped to a triple as follows:
	
	    specialVCUri rdfs:subClassOf *generalVCUri* .
	
	specialVCUri and *generalVCUri* are the RDF representations of the specializing and generalizing vertex classes, respectively.
	    Every generalization relationship between two edge classes is mapped to a triple as follows:
	
	    specialECUri rdfs:subPropertyOf *generalECUri* .
	
	specialECUri and *generalECUri* are the RDF representations of the specializing and generalizing edge classes, respectively.

resource [except for instances of rdfs:Class or rdf:Property]	vertex
⇒	Every resource which is neither an instance of rdfs:Class nor of rdf:Property is mapped to a vertex being instance of the vertex class corresponding to the resource's type. Resources without a specific type are mapped to instances of Vertex. The uriRef attribute takes the resource's URI reference as value.
⇐	Every vertex is mapped to a triple as follows: *uri* rdf:type *classUri* . *uri* is derived from the schema prefix, the qualified name of the vertex class, and the sequence number of the vertex. *classUri* is the RDF representation of the vertex class.

blank node	vertex
⇒	Every blank node is mapped to a vertex being instance of Vertex.
⇐	Instances of Vertex are mapped to blank nodes.

occurrence of an instance of rdf:Property [except for rdfs:subClassOf and rdfs:subPropertyOf, no literal object]	edge
⇒	Every occurrence of an instance of rdf:Property—except for rdfs:subClassOf and rdfs:subPropertyOf—in a triple whose object is not a literal is mapped to an edge being instance of the edge class corresponding to the rdf:Property instance. The edge connects the vertices corresponding to the triple's subject and object.
⇐	Every edge is mapped to a triple as follows: *startUri* *uri* *endUri* . *uri* is derived from the schema prefix and the qualified name of the edge class. *startUri* and *endUri* are the RDF representations of the edge's start and end vertices, respectively.

literal	attribute value
⇒	Every literal is mapped to an attribute value. The vertex which possesses the value corresponds to the subject of triple in which the literal occurs. The attribute corresponds to the instance of rdf:Property occurring in the triple.
⇐	Every attribute value is mapped to a triple as follows: *vertexUri* *attributeUri* *value* . The literal *value* is derived from the attribute value. *elementUri* is the RDF representation of the vertex possessing the attribute value. *attributeUri* corresponds to the rdf:Property instance representing the attribute.

RDF datatype	grUML domain
rdf:PlainLiteral	String
rdf:XMLLiteral	String
xsd:anyURI	String
xsd:base64Binary	String
xsd:boolean	**Boolean**
xsd:date	Record Date(century :Integer, year :Integer, month :Integer, day :Integer)
xsd:dateTime	Record DateTime(century: Integer, year :Integer, month :Integer, day :Integer, hour :Integer, minute :Integer, second :Double)
xsd:decimal	Double
xsd:double	**Double**
xsd:float	Double
xsd:gDay	Integer
xsd:gMonth	Integer
xsd:gMonthDay	Record GMonthDay(month :Integer, day :Integer)
xsd:gYear	Record GYear(century :Integer, year :Integer)
xsd:gYearMonth	Record GYearMonth(century :Integer, year :Integer, month :Integer)
xsd:hexBinary	String
xsd:int	**Integer**
xsd:long	**Long**
xsd:string	Enum
xsd:string	**String**
xsd:time	Record Time(hour :Integer, minute :Integer, second :Double)

Fig. 7. Mapping between RDF datatypes and grUML domains. **Boldface** denotes the mapping of choice when more than one possibility exists.

Example. Applying the schema-aware mapping approach on the following RDF triples results in the schema and TGraph in Figure 8. The original RDF graph can be recovered by executing the inverse transformation on the TGraph.

```
# URI: http://de.uni_koblenz.rdf/addrBook
@prefix rdf: <http://www.w3.org/1999/02/22-rdf-syntax-ns#>
@prefix rdfs: <http://www.w3.org/2000/01/rdf-schema#>
@prefix xsd: <http://www.w3.org/2001/XMLSchema#>
@prefix addressBook: <http://de.uni_koblenz.rdf/addrBook#>

addressBook:Entry rdf:type rdfs:Class .
addressBook:Person rdf:type rdfs:Class .

addressBook:Person rdfs:subClassOf addressBook:Entry .

addressBook:knows rdf:type rdf:Property .
addressBook:knows rdfs:domain addressBook:Entry .
addressBook:knows rdfs:range addressBook:Entry .
addressBook:name rdf:type rdf:Property .
addressBook:name rdfs:domain addressBook:Entry .
addressBook:age rdf:type rdf:Property .
addressBook:age rdfs:domain addressBook:Person .

addressBook:entry1 rdf:type addressBook:Person .
addressBook:entry1 addressBook:name "Lisa" .
addressBook:entry1 addressBook:age "28"^^xsd:int .
```

```
addressBook:entry2 rdf:type addressBook:Person .
addressBook:entry2 addressBook:name "Hugo" .
addressBook:entry2 addressBook:age "26"^^xsd:int .

addressBook:entry1 addressBook:knows addressBook:entry2 .
```

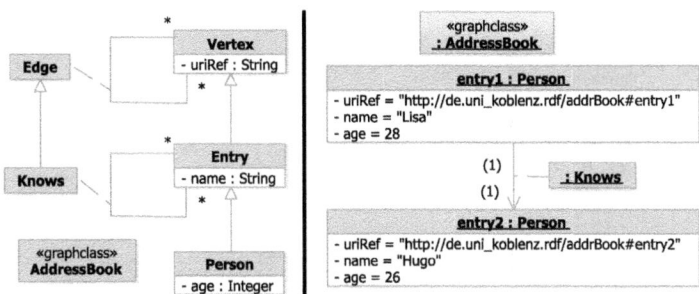

Fig. 8. Schema (left) and TGraph (right) as result of a schema aware mapping

4.2   Simple Mapping of RDF to TGraphs

The TGraph schema depicted in Figure 9 is suitable to instantiate TGraphs representing RDF graphs of arbitrary structure. Since the schema is directly created on the basis of the RDF concepts explained in Section 2.1, a detailed explanation of the schema is not included here. The mapping of RDF concepts to TGraph concepts is performed according to the table in Figure 10.

The instance-of hierarchy is flattened by treating rdf:type properties as instances of Arc with the appropriate value for the uriRef attribute. Furthermore, predicates are transformed to Arcs, which are distinguished first class elements instead of mere occurrences of instances of rdf:Property.

5   Transformation between SPARQL and GReQL

As it can be discerned from the descriptions of SPARQL and GReQL in Sections 2.2 and 3.2, both query languages provide various features which are not shared by the respective other language. Examples are the construction of RDF graphs by SPARQL CONSTRUCT queries and the usage of stand-alone GReQL RPEs. Consequently, SPARQL and GReQL are not mutually substitutable, i.e., it is not always possible to replace a SPARQL query by an equivalent GReQL query, and vice versa. Considering Figure 11 which provides an overview of the mapping between the different query parts in both languages, it is evident that the central subjects of such a mapping are SELECT and ASK queries in SPARQL and FWR expressions and existentially quantified expressions in GReQL.

For reasons of brevity, the direct transformation between ASK queries and existentially quantified expression is not considered. Instead, ASK queries are

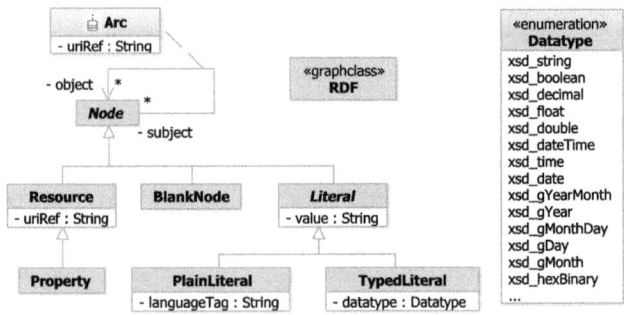

Fig. 9. TGraph schema for representing any RDF graph

RDF concept	TGraph concept	Comment
subject/object		
resource	instance of Resource/ instance of Property	The URI reference is mapped to the uriRef attribute. There is exactly one instance of Resource for a given URI reference. rdf:Property instances are mapped to Property instances.
blank node	instance of Blank-Node	There is exactly one instance of BlankNode for any blank node identifier in the RDF graph.
plain literal	instance of Plain-Literal	The value and language tag are mapped to the value and languageTag attributes, respectively. Occurrences of the same value in the RDF graph are mapped to the same instance.
typed literal	instance of Typed-Literal	The value and datatype are mapped to the value and datatype attributes, respectively. Occurrences of the same value in the RDF graph are mapped to the same instance.
predicate	instance of Arc	The URI reference is mapped to the uriRef attribute. There is exactly one instance of Arc for any predicate. If the rdf:Property instance does not exist as subject or object in the RDF graph, an additional, corresponding Property instance is created (needed for the SPARQL-to-GReQL transformation – Section 5.1.)

Fig. 10. Simple mapping from RDF to TGraph concepts

transformed to an equivalent union of FWR expressions where the resulting bag is checked for non-emptiness. If desired, this union can be straightforwardly rewritten as a quantified expression: the variables declared in the from parts are mapped to declarations in the quantified expression. The with parts go to the boolean expression after the @-sign, connected by logical or operators. Conversely, it is analogously assumed that existentially quantified expressions are first rewritten as FWR expression whose result is tested for non-emptiness.

Depending on whether schema-aware or simple mapping is used to transform RDF, the transformation steps to be applied on a SPARQL query differ slightly. It is analogously distinguished between *schema-aware* and *simple transformation* from SPARQL to GReQL. In the following, Section 5.1 explains schema-aware transformation in detail and shortly sketches the required modifications for simple transformation. In Section 5.2, the transformation from GReQL to SPARQL is described. Since the TGraph-to-RDF mapping is always schema-aware, no distinction between schema-aware and simple transformation is made here.

SPARQL query part		GReQL concept	Comment
prologue		—	no correspondence for base URI or prefixes in GReQL
query form	SELECT	report part of (a union of) FWR expression(s)	If UNION or OPTIONAL are used, it maps to the union of multiple FWR expressions instead of a single one. If DISTINCT or REDUCED are specified, reportSet is used.
	CONSTRUCT	—	no support in GReQL for building graphs
	ASK	existentially quantified expression	
	DESCRIBE	—	no support for returning descriptions specified in graphs in GReQL
dataset		—	no support for querying multiple graphs in GReQL
where clause	with SELECT query form	from and with parts of (a union of) FWR expression(s)	Variables in the where clause are mapped to declarations in the from part(s). Triple patterns and FILTERs are mapped to the with part(s).
	with ASK query form	declaration part and boolean expression in existentially quantified expression	Variables in the where clause are mapped to declarations in the declaration part of the quantified expression. Triple patterns and FILTERs are mapped to the boolean expression.
solution modifier		—	no support in GReQL for sorting or controlling the number of solutions
—		*all other expressions, e.g., RPEs*	no support in SPARQL for other GReQL concepts

Fig. 11. Mapping between SPARQL query parts and GReQL concepts

5.1 Transforming SPARQL to GReQL

Before presenting the transformation from SPARQL to GReQL in detail, some restrictions on SPARQL queries which are to be transformed are listed.

Restrictions on SPARQL queries for schema-aware mapping. Figure 11 already expresses some restrictions for the transformation of SPARQL queries, i.e., only SELECT and ASK queries operating on a single RDF graph and not making use of solution modifiers can be transformed. Provided that the queried RDF graph was transformed using schema-aware mapping, the restrictions on RDF graphs entail further requirements for the triple patterns in the where clause. The letters in the following list correspond to the letters of the origin in Section 4.1. However, restriction (f) on RDF graphs has no correspondence here.

(a) A variable acting as object in an rdf:type triple pattern must not occur as subject in other patterns, including rdfs:domain and rdfs:range patterns, for it is not possible to retrieve the start and end vertex classes of edge classes in GReQL. Further, such a variable must not occur as object in triple patterns without an rdf:type predicate and must not occur as predicate in any pattern.

(b) A variable acting as object in triple patterns where the range of the predicate is a datatype must not occur as object in patterns where the range of the predicate is not a datatype, and vice versa.

(c) A variable acting as predicate in triple patterns with literal objects must not occur as subject in rdfs:subClassOf patterns.

(d) A variable must not act both as predicate and as subject or object, except in rdfs:subPropertyOf, rdfs:domain, and rdfs:range triples.

(e) A variable must not act as object in more than one rdf:type triple pattern.

Note that if variables are used as predicates, they are potentially mapped to occurrences of rdf:type, rdfs:subClassOf, or rdfs:subPropertyOf. But it is not possible for a GReQL query to return instance-of and generalization relationships.

Schema-aware transformation. The steps for schema-aware transformation from SPARQL to GReQL are enumerated below. The following, rather artificial query including all needed SPARQL concepts serves as a running example.

```
PREFIX rdf: <http://www.w3.org/1999/02/22-rdf-syntax-ns#>
PREFIX addressBook: <http://de.uni_koblenz.rdf/addrBook#>

SELECT ?entry ?name1 ?name2
WHERE {
    {
        ?entry rdf:type addressBook:Person .
        ?entry addressBook:name ?name1 .
        ?entry addressBook:age ?age .
        FILTER ( ?age < 27 )
    } UNION {
        ?entry addressBook:name ?name1 .
        OPTIONAL {
            ?entry addressBook:knows ?entry2 .
            ?entry2 addressBook:name ?name2 .
        }
    }
}
```

1. By exploiting the commutativity and associativity of joins and unions as well as the distributivity of unions with joins and left joins, the where clause of the SPARQL query is brought into the *normal form* presented in [21]:

```
WHERE { P_1 UNION P_2 UNION ... UNION P_n }
```

Every $P_{i,i\in\{1..n\}}$ is a group graph pattern which is free of further UNION keywords. The example query already is in normal form.

2. For each P_i, a GReQL expression $expr_{i,i\in\{1..n\}}$ is formulated whose union is computed by first constructing a list of these expressions which is then passed to GReQL's union function:

```
union(list(expr₁, expr₂, ..., exprₙ)))
```

In the case of an ASK query, the above expression constitutes the parameter for an application of the isEmpty function, preceded by a logical not:

```
not isEmpty(union(list(expr₁, expr₂, ..., exprₙ)))
```

For the example SPARQL query, the resulting GReQL query looks as follows:

```
union(list(expr₁, expr₂))
```

3. The form of $expr_i$ is dependent on whether P_i contains OPTIONAL patterns. If not, $expr_i$ corresponds to a single FWR expression, hereafter denoted as FWR_i. Otherwise $expr_i$ is replaced by the union of multiple FWR expressions $FWR_{i_k,k\in\{1..r\}}$. The value of r depends on the exact structure of a P_i with OPTIONAL patterns. Each such P_i can be considered as a sequence of group graph patterns $P_{i_j,j\in\{1..m\}}$ which consist of triple patterns and FILTERs, but are free of further OPTIONAL keywords:

P_{i_1} **OPTIONAL** P_{i_2} **OPTIONAL** ... **OPTIONAL** P_{i_m}

As explained in Section 2.2, braces can be inserted to enforce computation orders other than from left to right. Abstracting from the numbering, it is assumed that the computation starts with P_{i_x} and P_{i_y}. According to the definition of left join, the result is the set of solution mappings fulfilling *both P_{i_x} and P_{i_y}*, unified with the set of solution mappings fulfilling P_{i_x}, but *not P_{i_y}*. It is possible to set up a table whose rows represent possible configurations of solution mapping fulfillment for each P_{i_j}, represented by the columns (see Figure 12a). The table cells denote whether the solution mappings have to fulfill the respective P_{i_j} (*true*) or not (*false*).

This initial table is extended along with the computation of P_i. The extension procedure depends on whether the next P_{i_j} to be considered is on the right-hand or on the left-hand side of the already computed part. In the former case, each row in the preceding table is combined with both *true* and *false* as values for the fulfillment of P. Under the assumption that the above table is extended by taking the left join of the result of P_{i_x} OPTIONAL P_{i_y} and some P_{i_z}, the resulting table can be taken from Figure 12b.

If the next P_{i_j} to be considered is on the left-hand side, its fulfillment is mandatory. So it is only needed to combine the boolean value *true* for that P_{i_j} with the configurations taken from the preceding table. Furthermore, if $P_{i_{j+1}}$ is not fulfilled, the fulfillment of the remaining group graph patterns is irrelevant as their fulfilling solution mappings would not be considered anyway. This is denoted by "—" in the respective table cells. Continuing

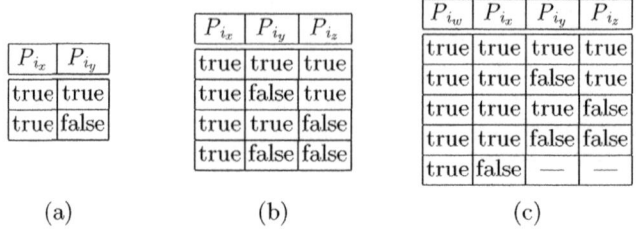

(a)   (b)   (c)

Fig. 12. Tables resulting from SPARQL queries with OPTIONAL patterns

with the above example, the table in Figure 12c reflects the left join of some P_{i_w} and the result of P_{i_x} OPTIONAL P_{i_y} OPTIONAL P_{i_z}:

Taking the final table with r rows for a P_i, each row k is mapped to an FWR expression $FWR_{i_{k,k \in \{1..r\}}}$. $expr_i$ corresponds to the union of all FWR_{i_k}:

```
union(list(FWR_i1, FWR_i2, ..., FWR_ir))
```

Applying this transformation step to the running example, the result is:

```
union(list(FWR1, FWR2_1, FWR2_2))
```

4. The from part of each FWR_i and FWR_{i_k} is determined as follows:
   (a) For each variable used in the where clause of P_i, a variable is declared in FWR_i.
   (b) For each variable used in the where clause of some P_{i_j} for which row k does not specify "—", a variable is declared in FWR_{i_k}.
   An exception is made for SPARQL variables occurring as objects in patterns whose predicates were transformed to an attribute. As attribute value access uses another mechanism in GReQL, such variables do not need to be mapped.

   For SPARQL variables used as objects in rdf:type triples, the domain of corresponding GReQL variables is the set of all vertex and edge classes, returned by the GReQL function types(). For variables used as subject or object, the domain is the set of the instances of the special vertex class V{Vertex}. Analogously, for SPARQL variables used as predicates, the domain is E{Edge} or the set of all attribute names, determined by the following GReQL expression. Here, the flatten function serves to convert the set of sets of attribute names for each vertex reported by the FWR expression to a single set containing all attribute names without duplicates.

```
flatten(from v:V{Vertex} reportSet attributeNames(v) end, true)
```

The from parts of the FWR expressions in the example look as follows:

```
FWR1:   from entry:V{Vertex}
FWR2_1: from entry, entry2:V{Vertex}
FWR2_2: from entry, entry2:V{Vertex}
```

5. The with part of each FWR_i and FWR_{i_k} is determined as follows:

(a) Due to join operation's commutativity and associativity, group graph patterns, possibly containing subordinate group patterns, can be safely split up into their components, i.e., triple patterns and FILTERs, so that every P_i has the following structure. Each F_i represents a FILTER and each T_i is a triple pattern:

FILTER (F_1) **FILTER** (F_2) ... **FILTER** (F_q) $T_1 \, . \, T_2 \, . \, ... \, T_p \, .$

All F_i and T_i are transformed to GReQL expressions connected by logical ands in the with part of FWR_i. Due to the multitude of filter expressions, a detailed description of their mapping to GReQL is out of scope here. Basically, GReQL provides most of the operators and functions offered by SPARQL. Missing functions can be added by extending GReQL's function library. Triple patterns are, in principle, mapped to path existence, forward, or backward vertex set expressions. As shown below, the transformation is to be performed in different ways, depending on the occurrences of RDF terms in a pattern. Special cases, e.g., rdf:type, rdfs:subClassOf, and rdfs:subPropertyOf triple patterns as well as literals as objects, result in the call of the corresponding GReQL functions hasType() and isA(), or in attribute value accesses, respectively.

To ease comprehension, self-explanatory identifiers directly referring to RDF terms and TGraph concepts are employed. Exceptions are PreE, referring to a predicate which is mapped to an edge, and PreA, denoting a predicate mapped to an attribute. Furthermore, GReQL let expressions are used to locally define variables whose scope is limited to the expression following the in keyword.

```
# SPARQL source
(1)  <Res> <PreE> <Res2> .              (14) ?var rdfs:subClassOf <Res> .
(2)  <Res> <PreA> Lit .                 (15) ?var rdfs:subPropertyOf <Res> .
(3)  <Res> rdf:type <Res2> .            (16) ?var1 <PreE> ?var2 .
(4)  <Res> rdfs:subClassOf <Res2> .     (17) ?var1 <PreA> ?var2 .
(5)  <Res> rdfs:subPropertyOf <Res2> .  (18) ?var1 rdf:type ?var2 .
(6)  <Res> <PreE> ?var .                (19) ?var1 rdfs:subClassOf ?var2 .
(7)  <Res> <PreA> ?var .                (20) ?var1 rdfs:subPropertyOf ?var2 .
(8)  <Res> rdf:type ?var .              (21) <Res> ?var <Res2> .
(9)  <Res> rdfs:subClassOf ?var .       (22) <Res> ?var Lit .
(10) <Res> rdfs:subPropertyOf ?var .    (23) ?var1 ?var2 .
(11) ?var <PreE> <Res> .                (24) ?var1 ?var2 <Res> .
(12) ?var <PreA> Lit .                  (25) ?var1 ?var2 Lit .
(13) ?var rdf:type <Res> .              (26) ?var1 ?var2 ?var3 .
```

```
// GReQL target
// for (7), (17), (23), (26), the attribute value is accessed in the report part (step 6)
(1)  let vertex := from v:V{Vertex} with v.uriRef = "Res" report v end
         in not isEmpty(vertex -->{PreE}&{@ thisVertex.uriRef = "Res2")
(2)  let vertex := from v:V{Vertex} with v.uriRef = "Res" report v end
         in hasAttribute(vertex, "preA") and vertex.attr = attrValue
(3)  let vertex := from v:V{Vertex} with v.uriRef = "Res" report v end
         in hasType(vertex, "Res2")
(4)  isA("Res", "Res2")
(5)  isA("Res", "Res2")
(6)  not isEmpty({@ thisVertex.uriRef = "Res"}&-->{PreE} var)
(7)  let vertex := from v:V{Vertex} with v.uriRef = "Res" report v end
         in hasAttribute(vertex, "preA")
```

```
(8)   let vertex := from v:V{Vertex} with v.uriRef = "Res" report v end
          in hasType(vertex, var)
(9)   isA("Res", var)
(10)  isA("Res", var)
(11)  not isEmpty(var -->{PreE}&{@ thisVertex.uriRef = "Res"})
(12)  hasAttribute(var, "preA") and var.preA = attrValue
(13)  hasType(var, "Res")
(14)  isA(var, "Res")
(15)  isA(var, "Res")
(16)  var1 -->{PreE} var2
(17)  hasAttribute(var, "preA")
(18)  hasType(var1, var2)
(19)  isA(var1, var2)
(20)  isA(var1, var2)
(21)  let vertex := from v:V{Vertex} with v.uriRef = "Res" report v end
          in (not isEmpty(vertex --var->&{@ thisVertex.uriRef = "Res2"})
              or hasType(vertex, "Res2"))
          or isA("Res", "Res2")
(22)  let vertex := from v:V{Vertex} with v.uriRef = "Res" report v end
          in hasAttribute(vertex, var) and getValue(vertex, var) = attrValue
(23)  let vertex := from v:V{Vertex} with v.uriRef = "Res" report v end
          in (vertex --var1-> var2 or hasType(vertex, var2))
              or hasAttribute(vertex, var2))
          or isA("Res", var2)
(24)  startVertex(var2) = var1 and endVertex(var2).uriRef = "Res"
          or hasType(var1, "Res") or isA(var1, "Res")
(25)  hasAttribute(var1, var2) and getValue(var1, var2) = attrValue
(26)  startVertex(var2) = var1 and endVertex(var2) = var3
          or hasType(var1, var3) or isA(var1, var3)
```

For reasons of brevity, the above GReQL expressions omit some required function parameter type checks. For example, if a SPARQL variable whose domain is the union of all graph element classes and all vertices of a specific class, it has to be tested whether the variable currently refers to a class before applying the isA function.

If a variable multiply occurs as object in triple patterns whose predicates are mapped to attributes, the equality of the attribute values has to be ensured by adding appropriate GReQL expressions.

(b) Each group graph pattern P_{i_j} for which row k of the corresponding table does not specify "—" is transformed to a group of GReQL expressions connected by logical ands in the with part of FWR_{i_k}, according to step 5a. In addition, for P_{i_j} where row k specifies $false$, the corresponding conjunction of expressions is negated by putting it into parentheses with a preceding logical not. Finally, the conjunctive GReQL expressions resulting from mapping each P_{i_j} are again interrelated by logical ands. Applying this step on the example results in the following with parts:

```
FWR₁:   with hasType(entry, "Person") and entry.age > 27
FWR₂₁:  with entry -->{Knows} entry2
FWR₂₂:  with not (entry -->{Knows} entry2)
```

6. For SELECT queries, the report part of each FWR_i and FWR_{i_k} is determined as follows. For ASK queries, it is only relevant that *something* is reported, with the concrete values being unimportant. For example, null can be specified after the report keyword in that case.

(a) For each variable specified after SELECT and used in the where clause of P_i, an expression involving the corresponding GReQL variable is reported by FWR_i. As SPARQL does not create bindings for variables specified after SELECT but not used in the where clause, null is specified in the report part in that case.

(b) For each variable specified after SELECT and used in the where clause of some P_{i_j} for which row k specifies *true*, an expression involving the corresponding GReQL variable is reported by FWR_{i_k}. If row k specifies *false* or "—" in all P_{i_j}, null is specified in the report part of FWR_{i_k}.

If the variable is used as predicate in SPARQL, the corresponding GReQL variable is reported itself. Otherwise, the uriRef attribute of the vertex represented by the GReQL variable is accessed and reported. An exception is made for variables occurring as objects in triple patterns where the predicate is mapped to a grUML attribute. Such variables are not included in the report part themselves, but the respective attribute value is accessed. If the variable is used as object in more than one pattern, it is irrelevant which value is accessed, as they are equal.

If DISTINCT or REDUCED are applied in the SPARQL query, report is replaced by reportSet, thus eliminating duplicate solutions.

The report parts of the FWR expressions in the example are:

```
FWR₁ : report entry.uriRef, entry.name, null
FWR₂₁ : report entry.uriRef, entry.name, entry2.name
FWR₂₂ : report entry.uriRef, entry.name, null
```

The whole GReQL expression resulting from performing all transformation steps on the initial SPARQL example is shown below.

```
union(list(
    from entry:V{Vertex}
    with hasType(entry, "Person")
        and entry.age > 27
    report entry.uriRef, entry.name, null end,
    from entry, entry2:V{Vertex}
    with entry -->{Knows} entry2
    report entry.uriRef, entry.name, entry2.name end,
    from entry, entry2:V{Vertex}
    with not (entry -->{Knows} entry2)
    report entry.uriRef, entry.name, null end
))
```

Simple transformation. If simple mapping was used to transform the RDF graph to be queried, SPARQL queries do not have to adhere to any restrictions. However, as sketched below, the transformation steps differ slightly with respect to the form of the FWR expressions. The numberings in the following list correspond to the transformation steps given above.

1–3. *no changes*

4. The domain of variables declared in the from part is V{Node} for all variables acting as subject or object but not as predicate. For variables only occurring as predicate, the domain is V{Property}. If a variable is

used both as subject or object and as predicate in the SPARQL query, two distinct variables have to be declared in GReQL, one with V{Node} and the other with V{Property} as domain.

5. Similar to schema-aware transformation of queries, triple patterns are basically mapped to path descriptions. The main differences are:
   - triple patterns with rdf:type, rdfs:subClassOf, and rdfs:subPropertyOf as predicates are transformed to path descriptions instead of function calls,
   - triple patterns with literal objects are transformed to path descriptions instead of attribute value accesses, and
   - predicates in triple patterns are transformed to edge restrictions checking for the value of the edges' uriRef attribute instead of their type.

6. The rules for deciding whether an expression involving a GReQL variable or null is included in the report part remain unchanged. But since a variable can represent vertices of different types with different attributes (uriRef, value, languageTag, datatype), the concrete type has to be checked using nested *conditional expressions* before the appropriate attribute value can be accessed. From the pair of GReQL variables originating from a single variable in a SPARQL query, only one has to be returned.

Applying simple transformation on the SPARQL example results in the following GReQL expression. To save space, only the first report part shows the whole expression for accessing the possible attribute values of a returned variable.

```
union(list(
    from entry, name, age:V{Node}
    with not isEmpty(entry -->{@ thisEdge.uriRef = "rdf:type"}
                     &{@ thisVertex.uriRef = "addressBook:Person"})
        and entry -->{@ thisEdge.uriRef = "addressBook:name"} name
        and entry -->{@ thisEdge.uriRef = "addressBook:age"} age
        and age.value < 27
    report hasType(entry, "Resource") ? entry.uriRef :
                hasType(entry, "PlainLiteral" ? entry.value+entry.languageTag :
                    hasType(entry, "TypedLiteral" ? entry.value+"^^"+entry.datatype :
                        hasType(entry, "BlankNode") ? entry : "error",
            ...
    end,
    from entry, name, entry2, name2:V{Node}
    with entry -->{@ thisEdge.uriRef = "addressBook:name"} name
        and entry -->{@ thisEdge.uriRef = "addressBook:knows"} entry2
        and entry2 -->{@ thisEdge.uriRef = "addressBook:name"} name2
    report ... end,
    from entry, name, entry2, name2:V{Node}
    with entry -->{@ thisEdge.uriRef = "addressBook:name"} name
        and not(entry -->{@ thisEdge.uriRef = "addressBook:knows"} entry2
            and entry2 -->{@ thisEdge.uriRef = "addressBook:name"} name2)
    report ... end
))
```

5.2   Transforming GReQL to SPARQL

Similar to SPARQL queries which are to be transformed to GReQL, GReQL queries also have to obey certain restrictions to be eligible for transformation to SPARQL. These restrictions are given in the following. Subsequently, the transformation from GReQL to SPARQL is explained stepwise.

Restrictions on GReQL queries. Relating to Figure 11, only FWR expressions and existentially quantified expressions can be transformed to SPARQL. Under the assumption that quantified expressions are rewritten as FWR expressions, there are, however, some further restrictions.

1. A variable declared in the from part may only have the set of instances of some vertex class vertices as domain.
2. The boolean expression in the with part may only consist of path existence expressions and particular function calls interconnected by the logical operators and/or. The evaluation order may be altered by parentheses.
3. RPEs must not employ edge path descriptions and iterations, as SPARQL does not contain comparable concepts.
4. The report part of may only include variables or attribute value accesses, as only those can be mapped to SPARQL variables.

The second restriction probably seems to be unnecessarily strong, considering the translation of triple patterns to GReQL expressions explained in step 5 of the SPARQL-to-GReQL transformation. However, since it is possible to reformulate let expressions as well as forward and backward vertex sets which are tested for non-emptiness as equivalent path existence expressions, the restriction adds much to the simplicity of the transformation.

Regarding function calls, it is out of scope for this work to review every single GReQL function for the existence of a SPARQL counterpart. But it can be said that many functions which take a boolean expression, a single graph element, or an attribute value as parameter and return a boolean value are transformable.

Transformation steps. Below, the GReQL-to-SPARQL transformation is described stepwise. The following, rather artificial query serves as running example.

```
from entry:V{AddressBookEntry}, person:V{Person}, address:V{Address}
with entry <--{Knows} person <-- <->{HasAddress} address and person.age < 27
report person, address.city end
```

1. Each RPE in the with part is brought into a *disjunctive normal form* by performing the following steps:
   (a) All simple path descriptions whose direction is irrelevant are replaced by an equivalent alternative path description:

   ```
   <-> ⇒ (--> | <--)
   ```

   (b) All optional path descriptions are replaced by alternatives as follows, with seq_1, seq_2, seq_3 denoting sequential path descriptions, i.e., concatenations of path descriptions:

   ```
   seq₁ [seq₂] seq₃ ⇒ (seq₁ seq₃ | seq₁ seq₂ seq₃)
   ```

   (c) All concatenations of alternative path descriptions are replaced by alternative sequential path descriptions:

$$
\begin{aligned}
&(seq_1 \mid seq_2 \mid \ldots \mid seq_p) \ (seq_{p+1} \mid seq_{p+2} \mid \ldots \mid seq_q) \\
\Rightarrow \ &(seq_1 \ seq_{p+1} \mid seq_1 \ seq_{p+2} \mid \ldots \mid seq_1 \ seq_q \\
&\quad \mid seq_2 \ seq_{p+1} \mid seq_2 \ seq_{p+2} \mid \ldots \mid seq_2 \ seq_q \\
&\quad \mid \ldots \\
&\quad \mid seq_p \ seq_{p+1} \mid seq_p \ seq_{p+2} \mid \ldots \mid seq_p \ seq_q)
\end{aligned}
$$

As result, every path existence expression will have the following structure:

$$
v \ (seq_1 \mid seq_2 \mid \ldots \mid seq_m) \ w
$$

Each $seq_{i,i \in \{1..m\}}$ denotes a sequential path description which possibly contains intermediate vertex path descriptions, but which is free of alternatives and options. v and w are arbitrary variables. Applying this transformation step to the with part of the example query results in:

```
with entry (<--{Knows} person <-- <--{HasAddress}
           | <--{Knows} person <-- -->{HasAddress}) address
      and person.age < 27
```

2. Each path existence expression with a path description in disjunctive normal form is replaced by a disjunction of multiple path existence descriptions:

$$
(v \ seq_1 \ w \ \textbf{or} \ v \ seq_2 \ w \ \textbf{or} \ \ldots \ \textbf{or} \ v \ seq_m \ w)
$$

The example's with part is modified as follows:

```
with (entry <--{Knows} person <-- <--{HasAddress} address
      or entry <--{Knows} person <-- -->{HasAddress} address)
      and person.age < 27
```

3. For each variable in the from part, a group graph pattern is created in the where clause. It contains a single rdf:type triple pattern with the variable as subject and the resource corresponding to the vertex class whose set of instances is the variable's domain as object. For each of the vertex class's subclasses, a similar, corresponding group graph pattern is added. All group graph patterns created for one variable in this way are interrelated by UNION.
   Transforming the from part of the example query results in:

```
{ { ?entry rdf:type <AddressBookEntry> } UNION { ?entry rdf:type <Person> } }
{ ?person rdf:type <Person> }
```

4. The with part is mapped to the where clause by transforming and operators to joins and or operators to UNION. Changes in the evaluation order by parentheses are reflected by an equivalent usage of group graph patterns.

5. Each path existence expression is transformed to triple patterns as follows. ?p, ?q, and ?r represent variables not occurring in another triple pattern. <PreE> is the rdf:Property instance corresponding to EdgeClass.

```
// GReQL source
(1)  v --> w      (3)  v <-> w        (5)  v -->{EdgeClass} x <-> w
(2)  v <-- w      (4)  v -->{EdgeClass} w   (6)  v <->{EdgeClass} <-- w
```

```
# SPARQL target
(1)   ?v ?p ?w .                  (4)   ?v <PreE> ?w .
(2)   ?w ?p ?v .                  (5)   ?v <PreE> ?x . ?x ?p ?w . ?w ?q ?x .
(3)   ?v ?p ?w . ?w ?q ?v .       (6)   ?v <PreE> ?r . ?v <PreE> ?r . ?w ?p ?r .
```

Transforming the path existence expressions in the example results in:

```
{
    {
        ?person <Knows> ?entry .
        ?r ?p ?person .
        ?address <hasAddress> ?r .
    } UNION {
        ?person <Knows> ?entry .
        ?r ?p ?person .
        ?r <hasAddress> ?address .
    }
}
```

6. Each function call is transformed to an equivalent FILTER. The choice of FILTER expressions for specific GReQL functions is not detailed here.

   If an attribute value is accessed as part of a function call, an additional triple pattern is inserted. It introduces a previously unused variable representing the literal the attribute value has been mapped to. Note that <PreA> is the rdf:Property instance corresponding to the attribute attr.

```
// GReQL source
v.attr = value
```

```
# SPARQL target
v <PreA> ?lit . FILTER(?lit = value)
```

Transforming the call of the leThan() function in the example – where its infix syntax (<) is used – results in the following excerpt:

```
?person <age> ?age . FILTER (?age < 27)
```

7. If the FWR expression resulted from rewriting an existentially quantified expression, the query form is ASK. Otherwise, the report part of the FWR expression is transformed to the SPARQL SELECT query form, followed by DISTINCT to eliminate duplicate solutions. For each variable specified after report, a SPARQL variable is specified. If an attribute value is accessed, an additional triple pattern is inserted in the where clause. This pattern introduces a previously unused variable representing the corresponding literal.

   The final SPARQL query resulting from the application of all transformation steps on the initial GReQL query example looks as follows:

```
SELECT DISTINCT ?person ?city
WHERE {
    {
        {
            ?entry rdf:type <AddressBookEntry>
        } UNION {
            ?entry rdf:type <Person>
        }
    }
```

```
{ ?person rdf:type <Person> }
{
    {
        ?person <Knows> ?entry .
        ?r ?p ?person .
        ?address <hasAddress> ?r .
    } UNION {
        ?person <Knows> ?entry .
        ?r ?p ?person .
        ?r <hasAddress> ?address .
    }
}
?person <age> ?age . FILTER (?age < 27)
?address <city> ?city .
}
```

6   Applications

In the following, possible applications of the bridging approach illustrate its usability in practice. While Section 6.1 makes use of the approach to retrieve traceability information recorded in ontologies, section 6.2 shortly discusses further potential applications.

6.1   Querying Traceability Information Represented in Ontologies

Traceability information plays a major role in various fields, including project management, quality assurance, and maintenance. But despite the multitude of different applications, most traceability retrieval problems, i.e., problems addressing the extraction of required traceability information, can be reduced to one of three basic patterns [17]: *existence*, *reachable entities*, and *slice*. In short, existence checks whether there exists a path of traceability relationships which conforms to a particular structure between any two artifacts out of two given sets of traced artifacts. Reachable entities is concerned with determining all artifacts reachable from any artifact out of a given set of traced artifacts, via paths with a specific structure. Slice is similar to reachable entities, with the difference that intermediate artifacts lying on the paths shall be included in the result.

Obviously, the structure of eligible paths between traced artifacts is to be specified for all three patterns. GReQL with its powerful regular path expressions is well-suited for describing such structures. Provided that traceability information is recorded in an ontology, for example according to the approach in [22], the bridging approach given in Section 5.2 allows users to pose their queries in GReQL instead of using the more cumbersome SPARQL.

As an example, consider Figure 13. It depicts an excerpt of the traceability information for a customer relationship management system, represented as RDF graph. The excerpt includes the architecture component a5 which is responsible for the billing functionality. a5 fulfils functional requirement fr26 and adheres to non-functional requirement nfr57. Both traceability relationships are represented as arcs. Furthermore, the graph contains traceability information on the stakeholders which originated or contributed to the requirements.

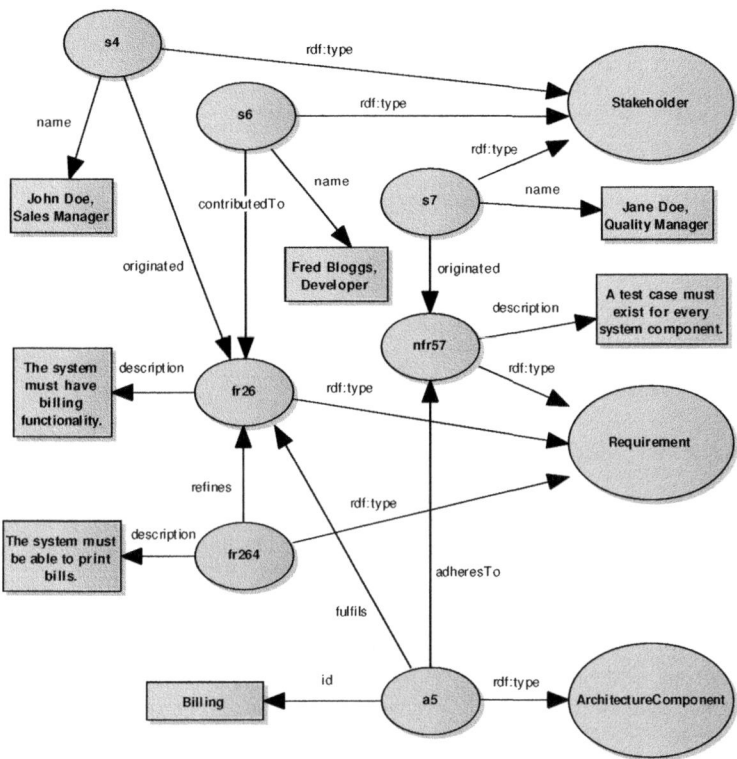

Fig. 13. Example traceability information recorded in an RDF graph

Assuming that a change is applied to the billing component which has a potential impact on the requirements, the names of the relevant stakeholders to be informed have to be determined. This corresponds to an application of the reachable entities traceability retrieval pattern. Since there are different traceability relationship types connecting architecture components to requirements and requirements to stakeholders, the query needs to consider all these possible path structures. Applying the bridging approach described in this paper, GReQL with its regular path expressions can be employed. The following example query accomplishes the retrieval task:

```
from archComp:V{ArchitectureComponent}, stakeholder:V{Stakeholder}
with archComp.id = "Billing"
     and archComp (-->{Fulfils} | -->{AdheresTo})
                  (<--{Originated} | <--{ContributedTo}) stakeholder
report stakeholder.name end
```

The transformation of the above GReQL query to SPARQL, according to Section 5, yields the following query:

```
SELECT DISTINCT ?name
WHERE {
    ?archComp rdf:type <ArchitectureComponent> .
    ?stakeholder rdf:type <Stakeholder> .
    ?archComp <id> ?id . FILTER regex(?id, "Billing")
    {
        {
            ?archComp <fulfils> ?r .
            ?stakeholder <originated> ?r .
        } UNION {
            ?archComp <fulfils> ?r .
            ?stakeholder <contributedTo> ?r .
        } UNION {
            ?archComp <adheresTo> ?r .
            ?stakeholder <originated> ?r .
        } UNION {
            ?archComp <adheresTo> ?r .
            ?stakeholder <contributedTo> ?r .
        }
    }
    ?stakeholder <name> ?name .
}
```

Considering the example, it becomes clear that the manual specification of suitable SPARQL queries for retrieving traceability information requires more effort than similar GReQL queries, especially as the path structures – reflected in a single regular path expression in GReQL – become more complex.

A different situation arises if refinement relationships between requirements have to be considered, e.g., if in Figure 13, the fulfils arc a5 would be connected to fr264 instead of fr26. Since such "refinement chains" between requirements can be of any length, the GReQL has to include an iterated path description:

```
from archComp:V{ArchitectureComponent}, stakeholder:V{Stakeholder}
with archComp.id = "Billing"
    and archComp (-->{Fulfils} | -->{AdheresTo}) -->{Refines}*
                    (<--{Originated} | <--{ContributedTo}) stakeholder
report stakeholder.name end
```

This GReQL query cannot be transformed to SPARQL, as the latter language does not offer a comparable concept. However, in this case, the RDF graph could be transformed to a corresponding TGraph using the mapping approach given in Section 4.

6.2   Other Applications

Using of the bridging approach for traceability information retrieval given in Section 6.1 illustrated that with regular path expressions, GReQL has a decisive advantage over SPARQL in fields relying on the analysis of complex structures of relationships between entities. Another such application where regular path expressions are useful is program analysis in software reengineering. If a software system's implementation is represented in a source code ontology, e.g., as advocated in [22], a similar approach as described in Section 6.1 could be taken for querying relationships between source code elements with GReQL. Suitable queries for analyses can be taken from the GUPRO project [16] (Generic Understanding of Programs), which employed TGraphs together with GReQL for program analysis, or from [23], where the computation of program slices [24] is reduced to queries.

In *ontology-driven software development* as pursued by the MOST project, software developers make use of both the semantic as well as the model-based technological spaces to benefit from services not available in the respective other space. These include consistency, satisfiability, and subsumption checking, for instance, in the semantic space and model transformation, merging, and differencing in the model-based space [25]. The usage of the bridging approach between query languages in this paper relieves developers of the need to manually translate queries when transforming ontologies to models or vice versa.

7   Related Work

Most work in the field of bridging the semantic and the model-based technological spaces has apparently been done with respect to bridges between modeling languages, ranging from transformation bridges, e.g., from MOF to RDF in [26], to approaches targeting at an integration of technological spaces. Examples for the latter are the *ODM (Ontology Definition Metamodel)* to capture the structure of ontologies in MOF-compliant models [27], and the partly ODM-based TwoUse approach which integrates UML class diagrams with OWL concepts, including an OCL dialect for formulating constraints on integrated models [28].

Concerning bridges between query languages, one existing approach is the transformation from SPARQL to *JDOQL (Java Data Objects Query Language)* implemented by the *SPOON (SParql to Object Oriented eNgine)* tool [29]. JDOQL is an implementation of the *OQL (Object Query Language)* standard for usage with object-oriented database systems and is integrated with the JDO API[6] which serves to make Java objects persistent. The approach does not consider transforming JDOQL queries to SPARQL.

Another approach, presented in [30], aims at bidirectional transformations between SPARQL and *HQL*, another dialect of OQL. Meta-models for both languages are provided so that queries can be represented as conforming models on which suitable model transformations are executed. Up to now, the authors did not extensively evaluate their work and compare it to similar approaches.

8   Conclusion and Future Work

This paper introduces a transformation-based bridging approach between two query languages of the *semantic* and *model-based technological spaces: SPARQL and GReQL*, respectively. Since the transformation of queries relies on the prior transformation of the queried data, an appropriate mapping approach between *RDF and TGraphs* is also discussed. TGraphs and GReQL are similar, but somewhat more expressive than the popular EMOF together with OCL.

---

[6] http://java.sun.com/jdo

The bridging approach comprises transformations in both directions between SPARQL and GReQL as well as between the underlying data structures, i.e., RDF graphs and TGraphs. The transformations do not preserve the semantics of RDF and the TGraph approach, most importantly concerning the open and closed world assumptions, respectively, as well as entailment in RDF. However, this issue is irrelevant for the transformation between the query languages.

It is shown that albeit being more general, RDF graphs are also less structured than TGraphs. Thus, while all TGraphs can be uniformly mapped to RDF, only transformations of RDF graphs with TGraph-like structures can be transformed using the reversible *schema-aware mapping* which makes use of all TGraph features. All other RDF graphs must be transformed to TGraphs conforming to a fixed schema by *simple mapping*. This results in an awkward modeling style and a reduced efficiency of prospective GReQL queries. Comparing SPARQL and GReQL, any SPARQL query can be transformed to GReQL, with the exception of queries constructing a graph or relying on external descriptions of RDF resources. Conversely, GReQL is an expression language with many expressions constituting valid queries and the query results being of many different types, so that only a very limited number of GReQL expressions can be transformed. Most importantly, there is no SPARQL equivalent for GReQL's *iterated path description*, involving the computation of transitive closure.

Besides its applicability in ontology-driven software development as advocated by the MOST project, i.e., to automatically transform queries between technological spaces together with the artifacts to be queried, the usefulness of the bridging approach is also shown by a more concrete application in the field of traceability. It is based on the insight that GReQL with its path expressions is a highly suitable language for querying traceability information. If this information is stored in ontologies, the GReQL-to-SPARQL transformation allows users to formulate their queries in GReQL instead of SPARQL, so that relationships between traced artifacts can be described much more concisely.

In addition to refining existing applications and looking for new ones, future work will keep track of further development of the SPARQL language and adapt the approach accordingly. Up to now, there exist various extensions which did not make it into the standard yet. Some of them include regular path expressions [31]. It remains to be seen which extensions are adopted into SPARQL version 1.1 and how they compare to GReQL with respect to expressivity and efficiency.

References

1. Kurtev, I., Bézivin, J., Aksit, M.: Technological Spaces: an Initial Appraisal. In: International Conference on Cooperative Information Systems, pp. 1–6 (2002)
2. Klyne, G., Carroll, J.J. (eds.): Resource Description Framework (RDF): Concepts and Abstract Syntax – W3C Recommendation (February 10, 2004), http://www.w3.org/TR/2004/REC-rdf-concepts-20040210
3. Motik, B., Patel-Schneider, P.F., Parsia, B. (eds.): OWL 2 Web Ontology Language Structural Specification and Functional-Style Syntax – W3C Recommendation (October 27, 2009), http://www.w3.org/TR/2009/REC-owl2-syntax-20091027

4. Object Management Group: Meta Object Facility (MOF) Core Specification, Version 2.0 (2006),
   http://www.omg.org/spec/MOF/2.0/PDF
5. Steinberg, D., Budinsky, F., Paternostro, M., Merks, E.: EMF: Eclipse Modeling Framework. Addison-Wesley Professional, Reading (2008)
6. Prud'hommeaux, E., Seaborne, A. (eds.): SPARQL Query Language for RDF – W3C Recommendation (January 15, 2008), http://www.w3.org/TR/2008/REC-rdf-sparql-query-20080115
7. Ebert, J., Bildhauer, D.: Reverse Engineering Using Graph Queries. In: Graph Transformations and Model Driven Engineering. Springer, Heidelberg (to appear, 2010)
8. Schwarz, H., Ebert, J., Lemcke, J., Rahmani, T., Zivkovic, S.: Using Expressive Traceability Relationships for Ensuring Consistent Process Model Refinement. In: Proc. of the 15th IEEE International Conference on Engineering of Complex Computer Systems (2010)
9. Object Management Group: Object Constraint Language, Version 2.2 (2010), http://www.omg.org/spec/OCL/2.2/PDF
10. Manola, F., Miller, E. (eds.): RDF Primer – W3C Recommendation (February 10, 2004), http://www.w3.org/TR/2004/REC-rdf-primer-20040210
11. Hayes, P. (ed.): RDF Semantics – W3C Recommendation (February 10, 2004), http://www.w3.org/TR/2004/REC-rdf-mt-20040210
12. Biron, P.V.B., Malhotra, A. (eds.): XML Schema Part 2: Datatypes – W3C Recommendation (May 2, 2001),
    http://www.w3.org/TR/2001/REC-xmlschema-2-20010502
13. Brickley, D., Guha, R. (eds.): RDF Vocabulary Description Language 1.0: RDF Schema – W3C Recommendation (February 10, 2004), http://www.w3.org/TR/2004/REC-rdf-schema-20040210
14. Malhotra, A., Melton, J., Walsh, N. (eds.): XQuery 1.0 and XPath 2.0 Functions and Operators – W3C Recommendation (January 23, 2007), http://www.w3.org/TR/2007/REC-xpath-functions-20070123
15. Ebert, J., Süttenbach, R., Uhe, I.: Meta-CASE in Practice: a Case for KOGGE. In: Olivé, À., Pastor, J.A. (eds.) CAiSE 1997. LNCS, vol. 1250, pp. 203–216. Springer, Heidelberg (1997)
16. Ebert, J., Kullbach, B., Riediger, V., Winter, A.: GUPRO. Generic Understanding of Programs - An Overview. Electronic Notes in Theoretical Computer Science, vol. 72(2) (2002),
    http://www.elsevier.nl/locate/entcs/volume72.html
17. Schwarz, H., Ebert, J., Winter, A.: Graph-based traceability: a comprehensive approach. Software and Systems Modeling (2009), http://springerlink.metapress.com/link.asp?id=109378
18. Bry, F.B., Furche, T., Linse, B.: The Perfect Match: RPL and RDF Rule Languages. In: Proceedings of the 3rd International Conference on Web Reasoning and Rule Systems (RR 2009), pp. 227–241 (2009)
19. Holland, D.A.: PQL Language Guide and Reference. Harvard School of Engineering and Applied Sciences (2009)
20. Berglund, A., Boag, S., Chamberlin, D., Fernández, M.F., Kay, M., Robie, J., Siméon, J. (eds.): XML Path Language (XPath) 2.0 – W3C Recommendation (January 23, 2007), http://www.w3.org/TR/2007/REC-xpath20-20070123

21. Pérez, J., Arenas, M., Gutierrez, C.: Semantics and Complexity of SPARQL. In: Cruz, I., Decker, S., Allemang, D., Preist, C., Schwabe, D., Mika, P., Uschold, M., Aroyo, L.M. (eds.) ISWC 2006. LNCS, vol. 4273, pp. 30–43. Springer, Heidelberg (2006)

22. Witte, R., Zhang, Y., Rilling, J.: Empowering Software Maintainers with Semantic Web Technologies. In: Franconi, E., Kifer, M., May, W. (eds.) ESWC 2007. LNCS, vol. 4519, pp. 37–52. Springer, Heidelberg (2007)

23. Schwarz, H.: Program Slicing - Ein dienstorientiertes Modell. Vdm Verlag Dr. Mller (2007), http://www.vdm-verlag.de/

24. Weiser, M.: Program Slicing. IEEE Transactions on Software Engineering 10(4), 352–357 (1984)

25. Ebert, J.: Software Engineering with Models and Ontologies. In: Proceedings of the 9th Ninth International Baltic Conference on Databases and Information Systems, Riga, Latvia (to appear, 2010)

26. Cranefield, S., Pan, J.: Bridging the gap between the model-driven architecture and ontology engineering. International Journal of Human-Computer Studies 65, 595–609 (2007)

27. Object Management Group: Ontology Definition Metamodel, Version 1.0 (2009), http://www.omg.org/spec/ODM/1.0/PDF

28. Silva Parreiras, F., Staab, S., Winter, A.: TwoUse: Integrating UML Models and OWL Ontologies. Technical Report 16/2007, Institut für Informatik, Universät Koblenz-Landau, Arbeitsberichte aus dem Fachbereich Informatik (2007)

29. Corno, W., Corcoglioniti, F., Celino, I., Valle, E.: Exposing Heterogeneous Data Sources as SPARQL Endpoints through an Object-Oriented Abstraction. In: Proc. of the 3rd Asian Semantic Web Conference on The Semantic Web (2008)

30. Hillairet, G., Bertrand, F., Lafaye, J.Y.: Rewriting Queries by Means of Model Transformations from SPARQL to OQL and Vice-Versa. In: Paige, R.F. (ed.) ICMT 2009. LNCS, vol. 5563, pp. 116–131. Springer, Heidelberg (2009)

31. Alkhateeb, F., Baget, J.F., Euzenat, J.: Extending SPARQL with regular expression patterns (for querying RDF). Journal of Web Semantics 7, 57–73 (2009)

Semantic Business Process Engineering*

Jens Lemcke, Tirdad Rahmani, and Andreas Friesen

SAP Research, CEC Karlsruhe
{jens.lemcke,tirdad.rahmani,andreas.friesen}@sap.com

Abstract. In this tutorial, we compare OWL-DL reasoning and Petri net analysis for validating refinement and grounding of business processes.

(1) Process refinement: Like in software engineering, the implementation of a business process involves different interacting roles, such as business expert, analyst, process architect, and developer. Each role designs and refines different abstractions of the process until it is sufficiently refined. It is important to verify that the process models of the different abstractions are consistent.

(2) Process grounding: A sufficiently refined process has to be mapped on existing IT systems. Ideally, IT systems consist of components with a semantic annotation of their behavior. The most specific process must respect all IT systems' behaviors. Formally capturing process semantics enables to check automatically for consistent process refinement and grounding.

The classic application of semantic techniques in the area of static models is well understood. The analysis of business processes deals with dynamics. Modeling dynamics is a challenge for current approaches of semantic Web services. We compare advantages and shortcomings of Petri net analysis and description logic (DL) reasoning for refinement and grounding validation.

1 Introduction

Besides the mere research of fundamental methods, the application of the methods in practice is relevant. In this tutorial, we apply semantic technologies to the engineering of business processes. That includes two fundamental activities:

1. Similar to software engineering, multiple differently abstract versions of the same business process are created by different human modelers that represent different roles involved in the business process engineering procedure. Common engineering procedures include feedback cycles. In a feedback cycle, previously created models are adapted. Understanding the implications of change on dependent models is a challenge for human modelers. That is referred to as refinement validation.
2. The business process is in the end mapped on software components. When deployed, the business process must not exceed the capabilities of the software components. That is called grounding validation. Grounding validation is cumbersome for a human.

---

* This work has been funded by the European Commission within the 7th Framework Programme project MOST no. ICT-2008-216691, http://most-project.eu.

U. Aßmann, A. Bartho, and C. Wende (Eds.): Reasoning Web 2010, LNCS 6325, pp. 161–181, 2010.

We do not look at data flow but concentrate on the flow of control. Grounding and refinement validation can be regarded as checking the consistency of behavioral models. Consistency checking involves capturing the semantics of the behaviors. Semantics of behavior is qualitatively different from semantics of structures. The main purpose of Semantic Web technologies like OWL-DL is capturing semantics of structures. For capturing the semantics of dynamics, different techniques like Petri nets are more common. In this tutorial, we compare an OWL-DL and a Petri net implementation of grounding and refinement validation. This material reveals particular challenges that the dynamic semantics of process models pose on translating the refinement and grounding problems to OWL-DL.

2   Background

In this section, we formally define the problems to be addressed, namely grounding and refinement. As a prerequisite, we start with formally defining process models and component models:

2.1   Definition of Processes

A business process is a collection of related, structured activities that produce a specific service or product for a particular customer or customers. A process model is an explicit formulation of the behaviors of a business process. Here, we use BPMN to graphically represent process models due to its wide industry adoption. Figure 1 shows four examples of BPMN diagrams.

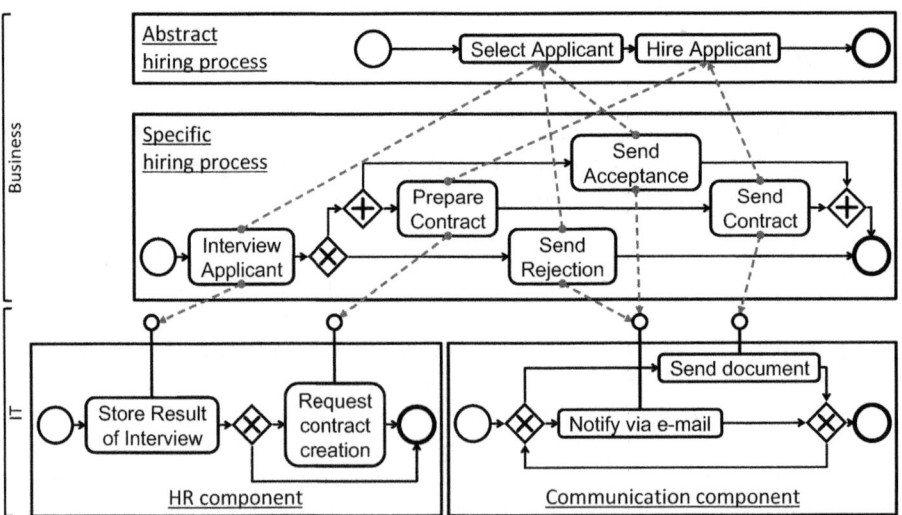

Fig. 1. Exemplary refined and grounded processes

Syntactically, we describe a process model $p \in \mathcal{P}$ as a directed graph $\Gamma_p = \langle V_p, E_p \rangle$. No two processes share the same vertices or edges: $\forall p, q \in \mathcal{P} : V_p \cap V_q = E_p \cap E_q = \emptyset$. The vertices are from the domain of vertices: $V_p \subseteq \mathcal{V}$. Vertices fall into events, activities, and gateways ($\Sigma, \mathcal{A}, \mathcal{G} \subseteq \mathcal{V}$), which are disjoint: $\Sigma \cap \mathcal{A} = \mathcal{A} \cap \mathcal{G} = \mathcal{G} \cap \Sigma = \emptyset$. We denote the subsets relevant to the specific process p by additional superscripts if reference would be ambiguous otherwise, for example, $A_p = \mathcal{A} \cap V_p$. We distinguish start and end events ($\Sigma^S, \Sigma^E \subseteq \Sigma$), which are disjoint: $\Sigma^S \cap \Sigma^E = \emptyset$. There is at most one start and one end event per process: $|\Sigma_p^S| = |\Sigma_p^E| = 1$. A gateway is either exclusive, denoted by $\circledast$, or parallel, denoted by $\circledcirc$: $\mathcal{G}^{\circledast}, \mathcal{G}^{\circledcirc} \subseteq \mathcal{G}, \mathcal{G}^{\circledast} \cap \mathcal{G}^{\circledcirc} = \emptyset$. We call a process p *normal* iff it does not contain parallel gateways: $G_p^{\circledcirc} = \emptyset$. The edge set is from the domain of edges ($E_p \subseteq \mathcal{E}$), which is a binary relation on $\mathcal{V}$. We define the predecessor and the successor functions of each $v_1 \in V_p$ as follows:

$$\mathrm{pre}(v_1) := \{\, v_2 \in V_p : (v_2, v_1) \in E_p \,\},$$
$$\mathrm{suc}(v_1) := \{\, v_3 \in V_p : (v_1, v_3) \in E_p \,\}.$$

A start event $s \in \Sigma_p^S$ (end event $e \in \Sigma_p^E$) does not have predecessors (successors): $|\mathrm{pre}(s)| = |\mathrm{suc}(e)| = 0$. Gateways and activities can have multiple predecessors and successors. The semantics is defined in the following.

The semantics of a process model p is based on its executions. An execution is a *proper* sequence of activities $a_i \in A_p$: $[a_1, a_2, \ldots, a_n]$. A proper sequence is obtained by simulating token flow through a process model: A token is associated to exactly one vertex or edge. Initially, there is exactly one token, associated to the start event. Tokens can be created and consumed following the rules below. Whenever a token is created in an activity, the activity is appended to the sequence. Exactly one of the following actions is performed at a time:

- For creating a token in an activity or an end event $v_1 \in A_p \cup \Sigma_p^E$, exactly one token must be consumed from exactly one incoming edge $(v_2, v_1) \in E_p$.
- Exactly one token must be removed from an activity or from a start event $v_1 \in A_p \cup \Sigma_p^S$ in order to create one token in every leaving edge $(v_1, v_2) \in E_p$.

- For creating a token in an exclusive gateway $g \in G_p^{\circledast}$, exactly one token must be consumed from exactly one incoming edge $(v, g) \in E_p$.
- Exactly one token must be removed from an exclusive gateway $g \in G_p^{\circledast}$ in order to create one token in exactly one leaving edge $(g, v) \in E_p$.

- For creating a token in a parallel gateway $g \in G_p^{\circledcirc}$, exactly one token must be consumed from every incoming edge $(v, g) \in E_p$.
- Exactly one token must be removed from a parallel gateway $g \in G_p^{\circledcirc}$ in order to create one token in each leaving edge $(g, v) \in E_p$.

If none of the above actions can be performed, simulation has ended. The result is a proper sequence of activities—an execution. Please note that by this definition, each execution is finite. It may be the case that the simulation described above never ends, but during an infinite run, no execution is created. Whenever the simulation reaches an end event, a finite execution is recorded. However, there may be an infinite number of executions for a process model. An infinite execution set is caused by at least one loop in the process model. The execution set of a process model p, denoted by ES_p, is the (possibly infinite) set of all proper sequences of the process model.

We only consider sound processes. According to the BPMN specification, unsoundness results from either deadlock or lack of synchronization. A process has a deadlock whenever no token coming from the start event can ever reach the end event considering the token flow semantics of BPMN. Lack of synchronization is given when a token reaches the end event although other tokens are still existing somewhere in the process. Examples are given in Figure 2.

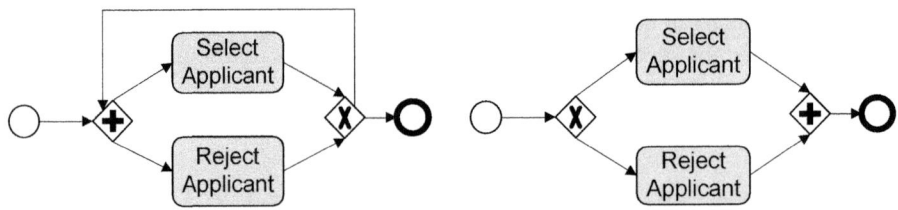

Fig. 2. Examples for lack of synchronization (left) and deadlock (right)

2.2 Definition of Components

A software component is a software element that conforms to a component model and can be independently deployed and composed without modification according to a composition standard [7]. Grounding can only be validated if the behavioral capabilities of software components are formally specified. Like proposed by semantic Web services research and by component-based software engineering, we assume that the behavioral capabilities of software components are explicitly modeled. A component model is an explicit formulation of the behaviors of a software component.

We define a new set of component models C with $\mathcal{P} \cap C = \emptyset$. Syntactically and semantically, a component model is a process model. The only difference is that operations ($\mathcal{O}$) are used instead of activities ($\mathcal{A}$). In practice, a component model has a passive role in that it defines the order of allowed operation invocations. In contrast, the process model plays an active role in that at run time it is expected to be pursued by either an orchestration engine or humans.

2.3 Definition of Grounding

Grounding is a function that assigns an operation to an activity: $\text{grnd} : \mathcal{A} \to \mathcal{O}$. Grounding validation involves checking completeness and correctness.

A process $p = \langle E_p, V_p \rangle$ is completely grounded iff $\text{grnd}()$ *is total for* A_p.

In other words, every activity of p must be grounded on some operation: $\forall a \in A_p$: $\text{grnd}(a)$ is defined. Please note that the operations a process grounds on may stem from different components. Checking completeness of grounding is a trivial syntactic check. In contrast, correctness of the grounding is more demanding.

A process $p = \langle E_p, V_p \rangle$ *is correctly grounded iff* p *grounds correctly on every component:* $\forall c \in \{ c \in C : a \in A_p \wedge \mathrm{grnd}(a) \in O_c \}$: p *grounds correctly on* c.

Our definition of correct grounding bases on the maximal execution set semantics used in the MIT process handbook project [16]. That article describes two subsumption semantics of processes based on the relations of the processes' execution sets:

A process p_1 *subsumes another process* p_2 *under maximal (minimal) execution set semantics iff* $ES_{p_2} \subseteq ES_{p_1}$ $(ES_{p_1} \subseteq ES_{p_2})$.

The execution sets of the exemplary processes in Figure 3 are: $ES_{p_1} = \{[a, b], [b, a]\}$, $ES_{p_2} = \{[a, b]\}$, and $ES_{p_3} = \{[a], [b]\}$. We base our grounding validation on the maximal execution set semantics because, informally, a subsumed process can neglect alternatives in executions, but has to follow the general behavior of the superior process. In the exemplary processes, $ES_{p_2} \subset ES_{p_1}$. In particular, p_2 preserves the execution $[a, b]$, but neglects the alternative execution $[b, a]$ of the subsuming process p_1. In contrast, p_3 does not even follow the general behavior of p_1 as $ES_{p_3} \cap ES_{p_1} = \emptyset$.

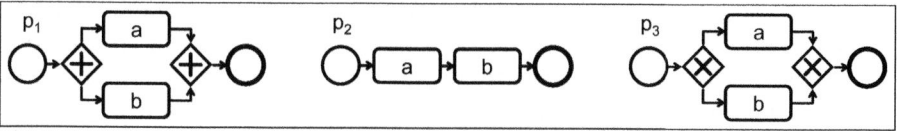

Fig. 3. Exemplary processes

Activities of a process and operations of a component are not equivalent. Consequently, formal correspondence between process and component must be established before their execution sets can be compared. We introduce two new notations for that purpose:

First, we connect activities that appear in an execution set with the operations they ground on. Therefore, we extend the signature of the grounding relation $\mathrm{grnd}()$ from activities to execution sets. That is again done in two steps: We start with extending the signature from activities to executions: $\mathrm{grnd}([a_1, a_2, \ldots, a_n]) = [o_1, o_2, \ldots, o_n]$, where $\mathrm{grnd}(a_i) = o_i$ for $i = 1 \ldots n$. We further extend the signature from executions to execution sets: Let p be a process, then $\mathrm{grnd}(ES_p) = \{ \mathrm{grnd}(e) : e \in ES_p \}$. Verbally put, $\mathrm{grnd}(ES_p)$ denotes the execution set that is yielded by substituting each activity of each execution of ES_p by the operation it grounds on. Please note that $\mathrm{grnd}(ES_p)$ contains in general operations from multiple different components.

Second, each execution of the process must be projected to only those activities that are grounded to the component we are comparing against. That is again done in two steps: First, we write $[o_1, o_2, \ldots, o_n]/c$ to denote the projection of execution $[o_1, o_2, \ldots, o_n]$ to only operations o_i of $c : o_i \in O_c$. Second, let ES be a set of executions. Then, $ES/c = \{ e/c : e \in ES \}$. Consequently, all executions of execution set ES/c consist only of operations of component c.

We can now define correct grounding:

A process p correctly grounds on component c iff $\mathrm{grnd}(ES_p)/c \subseteq ES_c$.

2.4   Definition of Refinement

Two processes participate in refinement: The (more) abstract and the (more) specific process. The abstract process is understood as control flow requirements to be observed in the more specific process. Formally, refinement is a function that assigns to an activity of a specific the activity of an abstract process it originates from: $\mathrm{orig} : \mathcal{A} \rightarrow \mathcal{A}$. Refinement is defined only between different processes: $\forall abs \in A_{p_{abs}}, spec \in A_{p_{spec}} :$ $\mathrm{orig}(spec) = abs \rightarrow p_{abs} \neq p_{spec}$. A process must refine at most one other process: $\forall a_1, a_2 \in A_{p_1}, a_3 \in A_{p_2} : \mathrm{orig}(a_1) = a_3 \rightarrow \mathrm{orig}(a_2) \in A_{p_2}$.

Refinement validation also involves checking completeness and correctness.

A process p_1 completely refines another process p_2 iff every activity of p_1 re-fines some activity of p_2 and every activity of p_2 is refined by some activity of p_1: $\forall a_1 \in A_{p_1} : \mathrm{orig}(a_1) \in A_{p_2}$ and $\forall a_2 \in A_{p_2} \exists a \in A_{p_1} : a_2 = \mathrm{orig}(a)$.

Again, checking refinement completeness is a trivial syntactic check. In contrast, check-ing refinement correctness is as complex as checking grounding correctness. Moreover, checking refinement correctness must allow for the decomposition of abstract activi-ties in the specific process. Therefore, we transform the abstract process before we apply a definition similar to the one for maximal execution set semantics in grounding correctness:

Let p be a process. Then, a decomposable process, denoted by p^D, is created from p as follows: For each activity $a \in A_p$, the connecting edges E_a, namely $\forall v_1, v_2 \in V_p :$ $(v_1, a), (a, v_2) \in E_a$, are replaced by a loop around a: The loop requires the addition of two new exclusive gateways, one new parallel gateway in the case of more than one outgoing edges, and respectively connected edges as displayed in Figure 4. Please note that the execution set of a decomposable process with at least one activity a is always infinite due to the loop around a.

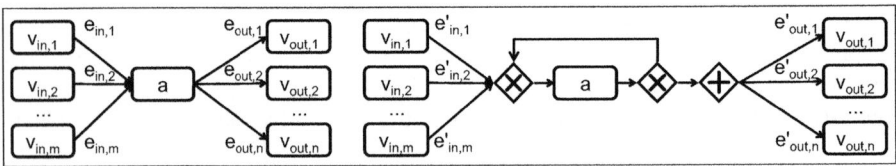

Fig. 4. Substitution of an activity for decomposition

As illustrated in Figure 5 on the facing page, the loop around A allows the spe-cific process to engage A arbitrarily often, namely by the refining activities $a1$ and $a2$, whenever and as long as the abstract process can execute A legally. The decomposable process enables us to determine correctness of refinement based on the subsumption of execution sets as follows.

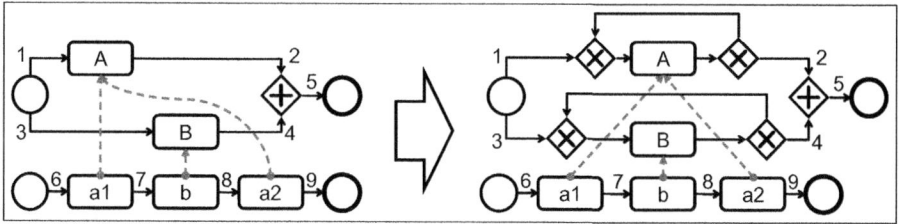

Fig. 5. Decomposition in exemplary abstract and specific process with refinement

The signature of orig() needs to be extended to executions and execution sets as we did for grnd() for the definition of correct grounding. Using the definition of a decomposable process and the extended signature of orig(), we can define correct refinement based on the idea of maximal execution set semantics:

A process p_1 correctly refines process p_2 (and p_2 is correctly refined by p_1) iff orig(ES_{p_1}) $\subseteq ES_{p_2^D}$.

2.5 Relevance of Refinement Validation

To assess the business value of refinement validation, we performed a paper-based multiple choice test with 13 experts from ontology and model-driven software development.[1] The test consists of two sets of 20 questions. Each question refers to three differently abstract models p_1, p_2, and p_3 of a real-life business process from [2], where p_1 refines p_2 and p_2 refines p_3. Each question has between two and four answer options where none or multiple answers can be correct. For the one set of 20 questions, denoted by S for "support", the result of the refinement validation is highlighted in the process models. The other set, denoted by N for "no support", has to be answered without support.

To exclude remembering effects, the models referred to in the two sets describe the same processes but with slightly different flaws. Furthermore, about one half of the test persons was given the support questions before the no-support questions and vice versa for the other half. We measured the number of correct and wrong answers per set (C_S, W_S, C_N, W_N) and the times taken for each set (t_S, t_N). The whole test took about 1.5 hours.

We calculate each person's quality of answers as $Q_x = C_x/(C_x + W_x)$ and productivity as $P_x = C_x/t_x$, where $x \in \{S, N\}$. In our test, the average answer quality without support was $\overline{Q_N} = 56\%$ and $\overline{Q_S} = 73\%$ with support. The productivity without support was $\overline{P_N} = 19.7\frac{\text{correct}}{\text{hour}}$ and $\overline{P_S} = 47.2\frac{\text{correct}}{\text{hour}}$ with support.

Although the whole averages already show an improvement through our work, we also calculate the improvement per person in quality as $QI = \frac{C_S/(C_S+W_S)}{C_N/(C_N+W_N)} - 1$ and in productivity as $PI = \frac{C_S/t_S}{C_N/t_N} - 1$. The numbers differ as not every person answered all questions per category. Our test reveals an average per-person improvement in quality of $\overline{QI} = 70\%$ and in productivity of $\overline{PI} = 378\%$.

---

[1] A preliminary online version of the test can be accessed at:
http://tiricompas.de/SurveySelection.php

3   Validating Grounding and Refinement with Petri Nets

A Petri net is a triple $\langle P, T, F \rangle$, where P is a finite set of places, T is a finite set of transitions, where $P \cap T = \emptyset$, and $F \subseteq (P \times T) \cup (T \times P)$ is a set of arcs (or the flow relation). Graphically, a place is denoted by an ellipse and a transition by a rectangle as in the lower part of Figure 6. We can define the set of input and output places of a transition t as $in(t) = \{ p \in P : (p, t) \in F \}$ and $out(t) = \{ p \in P : (t, p) \in F \}$. Similarly, the input and output transitions of a place p can be defined as $in(p) = \{ t \in T : (t, p) \in F \}$ and $out(p) = \{ t \in T : (p, t) \in F \}$. A place or transition $target \in P \cup T$ is reachable from another place or transition $source \in P \cup T$ if there is a directed path from source to target in the transitive closure of F. The state, or marking, of a Petri net is an assignment of tokens to places: $M \in P \to \mathbb{N}$. Graphically, we represent a token as a black dot inside a place.

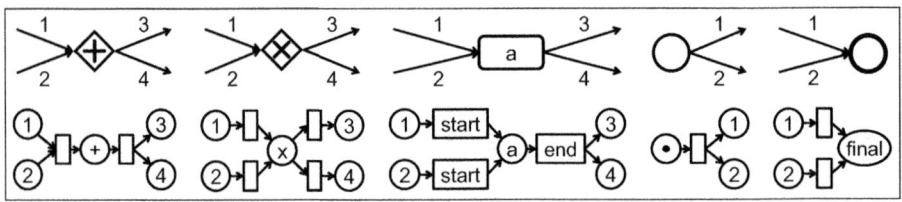

Fig. 6. Fragments of Petri nets corresponding to elements of process diagrams

Petri net executions are mathematically defined based on the firing of transitions: A transition can only fire if it is enabled. A transition is only enabled if all input places contain at least one token. When a transition fires, one token is consumed from each input place and one token is produced in each output place. An execution e of a Petri net is a sequence of markings: $e = (M_1, M_2, \ldots, M_n)$, where M_1 is the initial marking and M_{i+1} is the marking resulting from firing one transition in M_i for all $1 \le i \le n$. A Petri net has a deadlock if there is at least one marking M_i in any execution in which no transition is enabled. A Petri net is safe (or 1-bounded) if there is no marking in any execution where one place contains more than one token.

The idea of the grounding (refinement) validation using Petri nets is to simulate the synchronized executions of the process (specific process) and the component (abstract process). We construct a Petri net such that whenever the process (specific process) and the component (abstract process) get out of sync, the Petri net has a deadlock, and the grounding (refinement) is invalid and valid otherwise. As we need to check for subsumption of execution sets, the process (specific process) drives the simulation whereas the component (abstract process) tries to "follow" with steps corresponding to the grounding (refinement) function. We call the constructed Petri net the grounding (refinement) net.

The grounding (refinement) net is created in two steps detailed in the following sections. We use only the wording specific for grounding, but the procedure is the same for creating the refinement net except that we start from the decomposable abstract process as displayed in Figure 7 on the facing page.

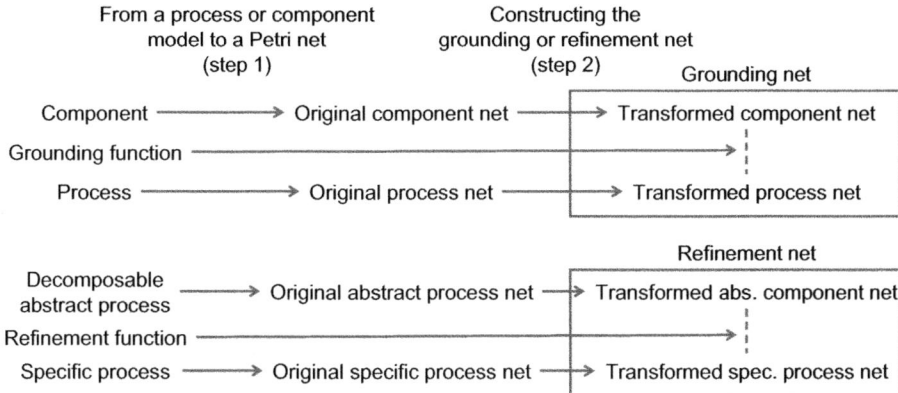

Fig. 7. Overview of creating the grounding and refinement nets

3.1   From a Process (Component) to a Original Process (Component) Net

First, we translate the process (component) to a Petri net. We call the resulting Petri net original process net (original component net).

In BPMN's token flow semantics, tokens are held by edges. Therefore, we create a new place $p \in P$ in the process net for each edge $e \in E$ in the process model and store the correspondence in the relation $E2P \subseteq E \times P$. Activities and gateways consume and produce tokens in BPMN's token flow semantics. As displayed in Figure 6 on the preceding page, we transform each kind of process (component) model vertices differently to Petri net structures due to their different behavior:

A **parallel gateway** $g^{\circledast}$ is used to synchronize and fork concurrent strands. Therefore, a new transition t is created for $g^{\circledast}$ and connected to the places corresponding to $g^{\circledast}$'s incoming and outgoing edges: A new arc $(p_{in}, t) \in F$ is created in the process net for each incoming edge $e_{in} \in \{(v_{in}, g^{\circledast}) \in E\}$ of the parallel gateway, where $(e_{in}, p_{in}) \in E2P$. A new arc $(t, p_{out}) \in F$ is created for each leaving edge $e_{out} \in (g^{\circledast}, v_{out}) \in E$, where $(e_{out}, p_{out}) \in E2P$.

An **exclusive gateway** $g^{\circledast}$ can split and merge alternative strands. Therefore, a new interim place is created for $g^{\circledast}$ which gets connected to a new transition per incoming and outgoing edge.

An **activity** a acts like an exclusive gateway for the incoming edges and like a parallel gateway for the leaving edges. Therefore, a new interim place—the activity place—is created for which we create a new input transition t_{start_i} per incoming edge, and one output transition t_{end} connecting all places corresponding to the leaving edges at once. It is to be noted that this transformation naturally separates the starting of an activity represented by the transitions t_{start_i}, called start transitions, and the ending of an activity represented by t_{end}, called end transition.

Every **start event** s^{S} is a concurrent starting point of the process and contains a token in BPMN's token flow semantics. Therefore, we create for the start event a new place p—the starting place—that contains a token in the initial marking M_1. A start event has no incoming edges and acts like a parallel gateway for the leaving edges. Therefore, we create a new transition and connect it to p as well as to all places corresponding to the leaving edges. The initial marking is the marking where one token is in every starting place and there are no other tokens in the net.

Every **end event** s^E consumes incoming tokens like an exclusive gateway. To ease the process net, we create a single place for all end events—the final place. The final place is connected to a new transition per incoming edge of each end event. A final marking is a marking where tokens are only in the final place.

The transformation of an exemplary process and a component connected by a grounding function is illustrated in Figure 8. It is also illustrated by dashed arrows that we can talk about grounded places in the process net similarly to grounded activities.

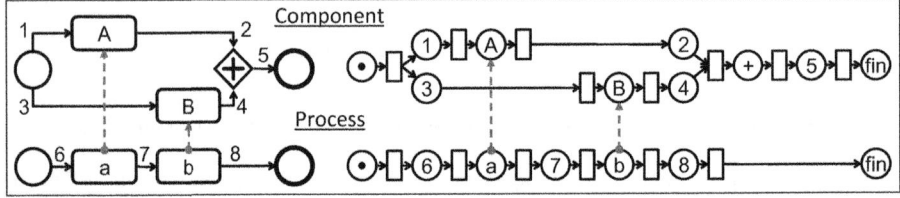

Fig. 8. Component and process net corresponding to exemplary component and process models

The computational complexity of constructing the process (component) net is linear in the number of nodes in the process (component) model because every node of a process (component) model is translated to a static Petri net structure as illustrated in Figure 6 on page 168.

3.2   Constructing the Grounding Net

The grounding net consists of slightly adapted versions of the process and the component net plus places, transitions, and arcs that model the grounding function. The grounding function is modeled in the grounding net such that the process plays an active role and the component follows passively. Additional places and arcs sync the executions of process and component. The detail is given in this section:

As the process should play an active role, starting a grounded activity a in the process is understood as a request to start the respective operation o in the component. Therefore, we create a request place $o!$ per operation o. That step's computational complexity is linear in the number of operations.

The request place $o!$ should contain a token iff o was requested by a. Therefore, we make every input transition of a an input transition of $o!$. Furthermore, o should only be executed if it was requested. Therefore, we make $o!$ an input place of every input transition of o. As an example, the new request place $B!$ becomes output (input) place of b's (B's) start transition in Figure 9 on the facing page. The worst-case complexity of that step is quadratic in the number of places and transitions: In the worst case, each place can be a request place and each transition may need to be connected to each request place.

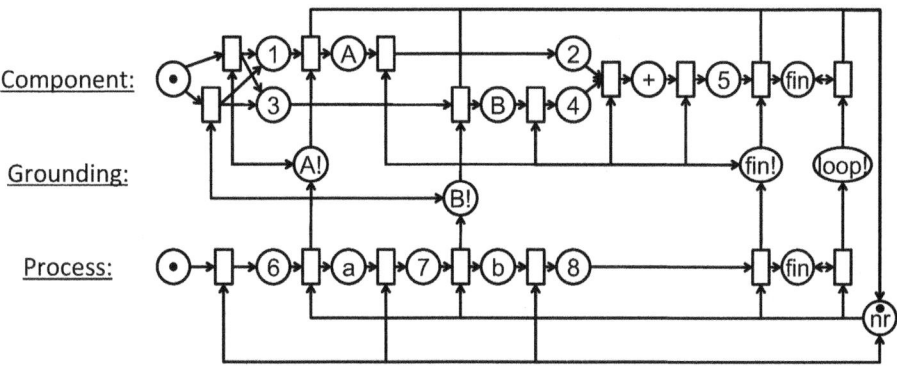

Fig. 9. Exemplary grounding net

We use deadlock as error condition for the grounding validation. Deadlock analysis is a standard task performed on Petri nets. Deadlock analysis returns a positive answer if at least one deadlock can be found in the net. Otherwise, the answer is negative. That decision implies that neither the component nor the process net must contain any deadlocks when they are sound. However, a sound process or component net may deadlock in a final marking. Therefore, we add to each final place p a new transition t—the final loop transition—and connect it by the two opposite arcs (p, t) and (t, p), graphically denoted by one arc with two heads. Now, neither the process nor the component model can cause a deadlock if it is sound—which we assume. To implement the request notion also for the process' (component's) loop transition $t^{loop}_{process}$ ($t^{loop}_{component}$), we create a loop request place $loop!$ and make it an output (input) transition of $t^{loop}_{process}$ ($t^{loop}_{component}$). As there is only one final place per model, that step's computational complexity is constant with respect to the size of the process (component) model.

As the component plays a passive role, it should never perform a step on its own. We already ensured that for the input transitions of operations by connecting them from the request places. For every remaining transition t of the component net, we substitute each t by a new transition t_o for each operation place o that is (syntactically) reachable from t without visiting other operation places and connect it as follows: We make t_o both an input and an output transition of the corresponding request place $o!$. The worst-case complexity of that step is quadratic in the number of places and transitions: In the worst case, each place is an operation place and each transition would be substituted by as many new transitions as there are operation places. The search for (syntactically) reachable places is also quadratic in the number of places and transitions because every transition could be connected to every operation place in the worst case.

Once an operation is requested, only the component should be able to perform a step. Therefore, we create a no-request place nr that should contain a token iff no request place contains a token. We make nr an input place of all transitions in the process net. Please note that this still allows for arbitrary steps of the process such that it can exert its active role. Specific transitions that are not connected to request places additionally become input transition to nr as the no-request place should contain a token iff no request place contains a token. In order to ensure that the process may perform a step as long as nothing was yet requested in the beginning of an execution, the no-request place initially contains a token. Furthermore, the process should also be able to perform a step after a requested operation was started—in other words, the request was completed. Therefore, every input transition of any operation place in the component net becomes an input transition of nr. That step has linear complexity in the number of specific transitions as in the worst case each transition has to be connected to the no-request place.

As we have seen, the whole transformation has a quadratic computational complexity.

3.3   Analyzing the Petri Net

We have reformulated the grounding (refinement) validation to a deadlock analysis problem in Petri nets.

Deadlock analysis of a Petri net is usually performed by constructing the Karp-Miller coverability tree [10]. The root is the initial marking. A branch is a transition. A branch connects a first marking with another marking that results from firing the transition in the first marking. A marking where no transition is enabled identifies a deadlock. The path from the root is the execution that causes the deadlock. Constructing the coverability tree also deals with loops and infinitely growing numbers of tokens in addition to what was said here. Further detail can be gathered from [10] or related textbooks.

Whether or not the process correctly grounds on the component can be determined by deadlock analysis using the coverability tree of the grounding net. In addition, the cause for an incorrect grounding can be determined from the execution causing the deadlock: The last transition of the execution is the start transition of the activity that would violate the component model. Removing all but the activity (operation) start transitions from the grounding net's execution yields the execution of the process (component) involved in the violation. That information may help the software engineer to understand and correct the process.

Petri nets can be classified based on structural properties. In particular, the Petri nets constructed by our method for sound process and component models are 1-safe. Deadlock detection for 1-safe Petri nets has PSpace-complete computational complexity [4, 3].

4   Validating Grounding and Refinement with OWL-DL

Grounding (refinement) can be validated using OWL-DL [1]. The idea of transforming the grounding (refinement) validation to OWL-DL bases on analyzing whether the sets of allowed predecessors and successors of activities and operations (abstract and specific activities) in concrete executions subsume each other. Unfortunately, allowed predecessors and successors of executions cannot be determined directly from a process model that contains parallel gateways. Therefore, a normal process model needs to be constructed before the corresponding OWL-DL ontology can be derived. Constructing a normal process model is done in two steps: First, the blocks of the process that are subject to parallel flow—called parallel blocks—are detected. Second, the process is normalized by substituting parallel blocks by equivalent blocks that only contain exclusive gateways—called exclusive blocks.

In the following sections, we describe the block detection of a process (component), the normalization to a normal process (component), and the construction of a grounding (refinement) ontology. The procedure is illustrated in Figure 10.

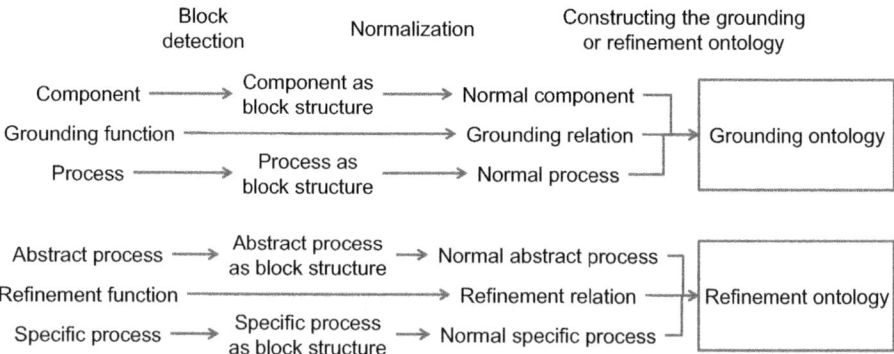

Fig. 10. Overview of creating the grounding and refinement ontologies

4.1 Block Detection

Block detection is performed on control flow graphs and SESE (single entry single exit) regions. For this purpose the following graph theoretical definitions taken from [9] have been used:

Definition 1. *A control flow graph G is a graph with distinguished nodes start and end such that every node occurs on some path from start to end. The start node has no predecessors and the end node has no successors.*

Definition 2. *A node x is said to dominate node y in a directed graph if every path from start to y includes x. A node x is said to postdominate a node y if every path from y to end includes x.*

Definition 3. *A SESE region in a graph G is an ordered edge pair (a, b) of distinct control flow edges a and b where*

1. *a dominates b,*
2. *b postdominates a, and*
3. *every cycle containing a also contains b and vice versa.*

Every process model can be seen as a control flow graph with the start event as the start node and the end event as the end node.

In this document SESE regions are called blocks and any algorithm for finding these blocks is called a block detection algorithm. Block detection is a prerequisite to find parallel execution blocks which we can then substitute by equivalent exclusive blocks. That is explained later.

There are also more sophisticated algorithms published in the recent years for block detection that are based on triconnected components in an undirected graph. The main work published regarding triconnected components has been carried out by [8]. Later, [6] enhanced and corrected that algorithm. [15] show how business process models can be decomposed to SESE regions in linear time.

In the following, we show examples for detected block types in some exemplary refinement scenarios. Although all types of blocks can be detected during block detection, normalization is restricted to structured blocks.

Definition 4. *A structured block is a SESE region that has at most two nodes with a higher indegree or outdegree than one. All other nodes in that block have indegree and outdegree equal to one.*

The simplest SESE region consists of a single activity. The activity is in that case the entry and exit node at the same time. A single activity forms the simplest block. The abstract process in Figure 1 on page 162 shows a graph with only simple blocks.

To ease the interpretation of block structures, we require that all gateways of the process must be either opening or closing gateways: $G^O, G^C \subseteq G$. An opening (closing) gateway $g \in G^O$ ($g \in G^C$) has exactly one predecessor (successor): $|\mathrm{pre}(g)| = 1$ ($|\mathrm{suc}(g)| = 1$). A gateway which is neither opening nor closing is called complex. Every complex gateway can be converted to exactly one closing and one opening gateway as shown in Figure 11 without changing the process flow.

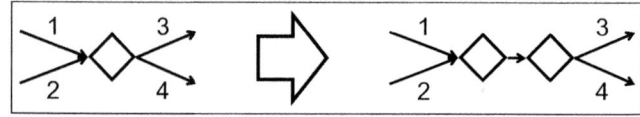

Fig. 11. Splitting complex gateways

We simplify process models further by substituting all activities with multiple predecessor or successors by two gateways and an equivalent activity with exactly one predecessor and successor as shown in Figure 12. We call a process model without complex gateways and without activities having multiple predecessors or successors a simple process.

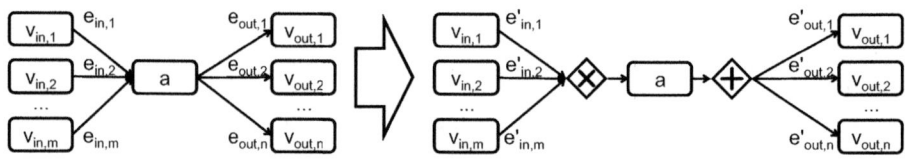

Fig. 12. Simplifying an activity

When we consider simple process models in the BPMN language, entry and exit nodes of structured, non-simple SESE regions can only be exclusive or parallel gateways since only gateways are allowed to have an in- or out degree greater than one.

For that reason, one can use the terms structured exclusive block or structured parallel block to refer to a structured block which has the same type of gateways as the entry and the exit node. Different types of gateways for structured blocks would lead to unsound BPMN diagrams, which we do not consider.

The definition of structured blocks implies that structured blocks can be decomposed into disjoint branches. Therefore, the process from Figure 13 can be written as $[a_1, \circledast([a_2], [b_1, b_2]), b_3]$. In that expression, brackets $[\dots]$ denote a branch consisting of sequential elements and $\circledast(\dots)$ denotes a structured parallel block. In the same way, we can write $\circledast(\dots)$ to denote an exclusive block.

Process to be transformed

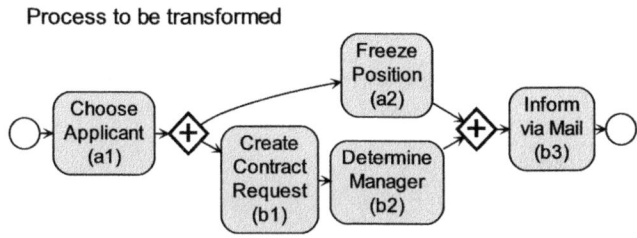

Fig. 13. Exemplary process with parallel block

Figure 14 shows a normal process model containing nested and unnested, non-simple blocks. The blocks are exclusive and structured and some contain loops.

Specific with Blocks

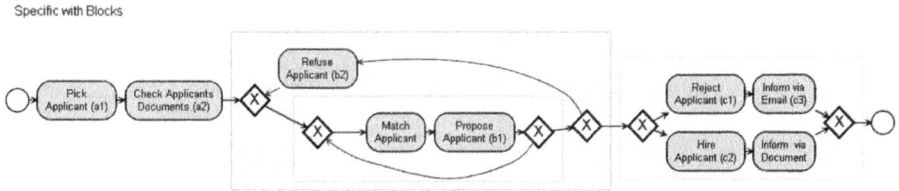

Fig. 14. Exemplary structured, exclusive blocks containing loops

The exemplary non-normal process model shown in Figure 15 on the next page illustrates the nesting of exclusive blocks in parallel blocks. All blocks are structured.

If a block is not structured, we call it unstructured. Figure 16 on the following page shows an unstructured parallel block. This block is not structured since it includes nodes (here: parallel gateways) that have an in- or out degree greater than one.

4.2 Normalization

The strategy for handling parallel blocks is to generate exclusive gateways to represent the executions based on predecessor and successor relations.

The normalization proposed by [12] starts from the nested block structure. The procedure restricts parallel blocks to be structured and must not contain further non-simple blocks. The procedure uses an extended notation for

Component1 with Blocks

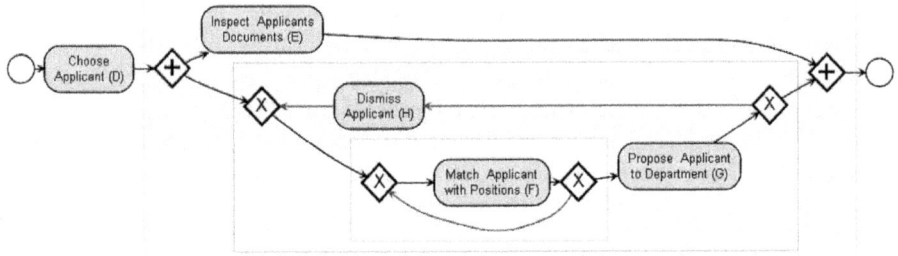

Fig. 15. Exemplary structured, nested exclusive and parallel blocks containing loops

Component2 with Blocks

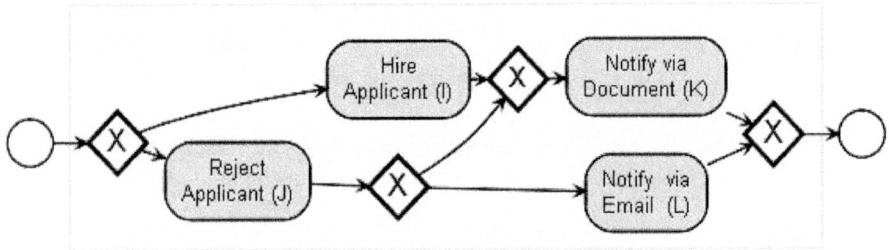

Fig. 16. Exemplary unstructured block

branches: $[a_1, \ldots, a_n|S]$ is a branch that starts with the sequential activities $a_1, \ldots, a_n$ followed by the sequence S. In the proposed normalization procedure, a parallel block $\circledast([a_1|S_1], \ldots, [a_n|S_n])$ is reduced to an exclusive block $\circledast([a_1, \circledast(S_1, [a_2|S_2]), \ldots, [a_n|S_n])], \ldots, [a_n|\circledast([a_1|S_1], \ldots, [a_{n-1}|S_{n-1}], S_n)])$.

Due to the drawback that only structured blocks without loops can be normalized, [13] proposed an alternative normalization: Given a process without complex gateways, the normalized process can be obtained as follows:

1. Repeatedly replace each opening parallel gateway g by an opening exclusive gateway e. For each $v \in \mathrm{suc}(e)$, construct a new opening parallel gateway g' such that $\mathrm{pre}(g') = v$, $\mathrm{suc}(g') = \mathrm{suc}(v) \cup \mathrm{suc}(e) \setminus \{v\}$ and then make $\mathrm{suc}(v) = g'$.
2. Remove all the edges from an opening parallel gateway to a closing parallel gateway.
3. If an opening gateway has only one successor, remove the gateway.
4. If a closing gateway has only one predecessor, remove the gateway.

In step 1, direct successors of parallel gateways are "pulled out" of the gateway. In the procedure, a parallel gateway with n successors is transformed to n parallel gateways with n successors, but one of the following sequences is shortened by one successor. Due to the finite length of the sequences, we know that the replacement always terminates. Step 2 then reduces the number of successors for the remaining parallel gateways by removing "empty" edges. Steps 3 and 4 finally remove the gateways. When a gateway is removed, its predecessors and successors are directly connected. It can easily be shown that the procedure always results in a normal process. An example of normalization of Figure 13 on page 175 can be seen in Figure 17. The abstract process previously presented in Figure 1 on page 162 did not change after normalization in contrast to the specific process previously presented in Figure 13 on page 175.

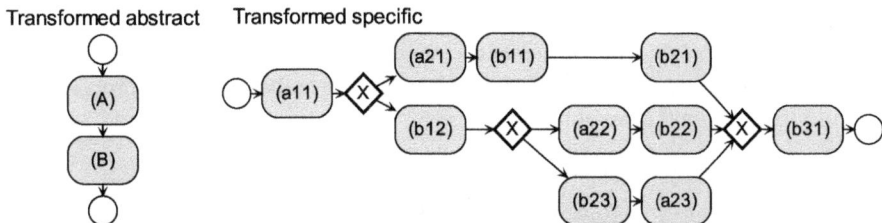

Fig. 17. Processes after normalization

The improved normalization deals with structured and unstructured blocks but still cannot handle loops. Loop-handling normalization is subject of future work. Both versions of the normalization have factorial complexity: $\mathcal{O}(n!)$.

As a consequence of the improved normalization, some operations (activities) of the original component (process) appear multiple times in the normalized component (process). Therefore, the grounding (refinement) function has to be adapted and, due to the multiplication of operations (abstract activities), it becomes a grounding (refinement) relation.

4.3   Constructing the Grounding and Refinement Ontologies

After the normalization of parallel to exclusive blocks, an ontology TBox can be created from the pair of process and one component (specific and abstract process). The precise rules are described in [12]. To give an overview, the construction follows some guiding principles:

1. Operations (activities) are represented as concepts.
2. Ordering and composition relations are represented as properties $from$ and to.
3. The component (abstract process) is represented by universal restriction.
4. The (specific) process is represented by existential restriction.
5. Uniqueness is represented by disjointness.

In particular, the creation procedure of the refinement ontology from the pair (s, a), where s is the specific and a the abstract process, undergoes four phases:

1. **Description of the abstract process.** Every abstract activity is related to its successors and predecessors. In addition, every abstract activity is related to itself. That

implements that specific activities originating from the same abstract activity can be execute as long as the abstract activity could be executed legally. That was also the purpose of the loop around an abstract activity in the decomposable process in the specification of refinement shown in Figure 5 on page 167. As an example, we list the relations for a process consisting of the sequential activities $[A, B]$ similar to the abstract hiring process in Figure 1 on page 162:

$$\text{Start} \sqsubseteq \forall \text{to. A} \qquad\qquad A \sqsubseteq \forall \text{from. (Start} \sqcup A)$$
$$A \sqsubseteq \forall \text{to. (A} \sqcup B) \qquad\qquad B \sqsubseteq \forall \text{from. (A} \sqcup B)$$
$$B \sqsubseteq \forall \text{to. (B} \sqcup \text{End)} \qquad\qquad \text{End} \sqsubseteq \forall \text{from. B}$$

2. **Description of the specific process.** Every specific activity is related to its successors and predecessors. As an example, we list some relations for the process shown in Figure 17 on the previous page:

$$\text{Start} \sqsubseteq \exists \text{to.a11} \qquad\qquad \text{a11} \sqsubseteq \exists \text{from.Start}$$
$$\text{a11} \sqsubseteq (\exists \text{to.a21}) \sqcap (\exists \text{to.b12}) \qquad\qquad \text{a21} \sqsubseteq \exists \text{from.a11}$$
$$\text{a21} \sqsubseteq \exists \text{to.b11} \qquad\qquad \text{b12} \sqsubseteq \exists \text{from.a11}$$
$$\text{b12} \sqsubseteq (\exists \text{to.a22}) \sqcap (\exists \text{to.b23}) \qquad\qquad \text{b11} \sqsubseteq \exists \text{from.a21}$$

3. **Expressing the refinement relation.** Specific and abstract activities are related to each other according to the refinement relation. For example: $a11 \sqsubseteq A$, $a21 \sqsubseteq A$, $a22 \sqsubseteq A$, $b11 \sqsubseteq B$, $b12 \sqsubseteq B$.
4. **Asserting uniqueness.** All abstract activities are disjoint. For example: disjoint $(\text{Start}, A, B, \text{End})$. Furthermore, all specific activities originating from the same abstract activity are disjoint. For example: disjoint $(a11, a21, a22)$ and disjoint $(b11, b12, b23, \ldots)$.

The grounding ontology is constructed the same way as the refinement ontology except that a grounding ontology is created for each pair of a process p and one component c_i individually. Before construction, p is projected to c_i. That means that every activity a in p that is not grounded to an operation in c_i is removed from p. Thereby, the incoming and outgoing edges to and from a are mutually connected as displayed in Figure 18 on the facing page. We denote the projected process by p'. The grounding ontology is then created similarly to the refinement ontology, but from the pair (p', c_i). As the only exception, operations are not put in relation to each other in phase 1 above as there is no decomposition in refinement validation.

The created ontology is fed to a description logics reasoner. The reasoner concludes a concept to be inconsistent if the corresponding activity is involved in a grounding or refinement violation. More detail and proofs are given by [13].

The complexity of description logics can be classified by the types of constructs they contain. The constructs created by the described procedure result in an $\mathcal{ALC}$ logic with acyclic TBox.[2] The reasoning complexity for that type of OWL-DL ontology is PSpace-complete.

---

[2] A configurator of description logics is available at:
http://www.cs.man.ac.uk/~ezolin/dl/.

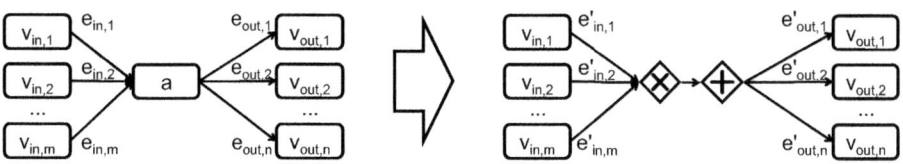

Fig. 18. Removing an action for projection

5 Conclusion

In this tutorial, we reviewed two approaches for grounding and refinement validation: once with Petri nets and once with OWL-DL. Clearly, both approaches commonly support the business process engineer during their tedious, manual, error-prone task of maintaining consistency between different abstraction level of process models and their grounding to components. Both approaches differ in their approach to the problem.

Petri nets were developed to express dynamic semantics. Therefore, Petri nets are by nature as rich as process and component models. Consequently, the grounding and refinement validation problems can be expressed rather directly using Petri nets.

In contrast, OWL-DL was developed to express static semantics. Therefore, only a part of the constructs of process and component models can be expressed directly in an ontology. Major difficulties result from parallel flow. Consequently, process and component models have to be normalized before an ontology can be created. The normalization is not straight-forward as the different approaches of [13, 12] have shown. Still, even the more advanced normalization cannot handle all kinds of process and component models.

The specific differences of both approaches are sketched in Table 1.

Table 1. Comparison

Category	Petri nets	OWL-DL
Model complexity	No restrictions (structured and unstructured blocks with arbitrary nesting)	No non-simple blocks nested in parallel blocks
Explanation	First error	All errors
Transformation complexity	Quadratic	Factorial
Analysis complexity	PSpace-complete	PSpace-complete

The advantage of the Petri net approach is that process and component models as well as the grounding and refinement function can be naturally translated to the formalism in a straight-forward way. The drawback of the Petri net solution is that only the first grounding or refinement violation can be found by deadlock detection.

The drawback of the Petri net solution is the advantage of the OWL-DL solution: All violations are immediately found. On the other hand, the solution for OWL-DL involves a complex graph transformation which can currently not deal with parallel blocks containing loop blocks.

From the complexity perspective, deadlock detection in Petri nets is exponential (PSpace-complete). The transformation required in the OWL-DL solution has factorial complexity ($\mathcal{O}(n) = n!$) which is worse than exponential ($\mathcal{O}(n) = e^n$). However, our experiments summarized in Table 2 show that in the OWL-DL solution, the transformation takes magnitudes of time less than the actual DL reasoning. That suggests that proper optimization techniques, such as approximation technology discussed by [11, 14], has the potential to improve the performance of the OWL-DL solution significantly by reducing the reasoning time.

Table 2. Statistics for OWL-DL-based refinement validation of 1239 generated abstract and specific processes

	Generic Activities	Specific Activities	Total Activities	Transf. Time	OWL-DL Axioms	DL Reasoning Time
Average	5.79	17.4	23.2	4ms	154	2.8s
Maximum	30	53	69	0.4s	1159	3.4min

From the comparison, we conclude that both approaches have advantages and challenges. Therefore, the area of semantic Web technologies can benefit from results found in related research areas such as from system dynamics and vice versa to improve existing algorithms and enter new application areas such as business process engineering.

References

Baader, F., Calvanese, D., McGuinness, D.L., Nardi, D., Patel-Schneider, P.F. (eds.): The Description Logic Handbook: Theory, Implementation, and Applications. Cambridge University Press, Cambridge (2003) ISBN 0-521-78176-0

Curran, T.A., Ladd, T., Ladd, A.: SAP R/3 Business Blueprint: Understanding Enterprise Supply Chain Management, 2nd edn. Prentice Hall International, Englewood Cliffs (1999)

Esparza, J.: Decidability and complexity of petri net problems - an introduction. In: Reisig, W., Rozenberg, G. (eds.) APN 1998. LNCS, vol. 1491, pp. 374–428. Springer, Heidelberg (1998) ISBN 3-540-65306-6

Esparza, J., Nielsen, M.: Decidability issues for petri nets - a survey. Bulletin of the EATCS 52, 244–262 (1994)

Grau, B.C., Horrocks, I., Motik, B., Sattler, U. (eds.): Proceedings of the DL Home 22nd International Workshop on Description Logics (DL 2009), CEUR Workshop Proceedings, Oxford, UK, July 27-30, vol. 477. CEUR-WS.org (2009)

Gutwenger, C., Mutzel, P.: A linear time implementation of spqr-trees. In: Marks, J. (ed.) GD 2000. LNCS, vol. 1984, pp. 77–90. Springer, Heidelberg (2001)

Heineman, G.T., Councill, W.T.: Component-Based Software Engineering: Putting the Pieces Together, 1st edn. Addison-Wesley Professional, Reading (2001)

Hopcroft, J.E., Tarjan, R.E.: Dividing a graph into triconnected components. SIAM J. Comput. 2(3), 135–158 (1973)

Johnson, R., Pearson, D., Pingali, K.: The program structure tree: Computing control regions in linear time, pp. 171–185. ACM Press, New York (1994)

Karp, R.M., Miller, R.E.: Parallel program schemata. Journal of Computer and System Sciences 3(2), 147–195 (1969)

Pan, J.Z., Thomas, E., Zhao, Y.: Completeness guaranteed approximations for owl-dl query answering. In: Grau, et al. (eds.) (2009)

Ren, Y., Gröner, G., Lemcke, J., Rahmani, T., Friesen, A., et al.: Validating process refinement with ontologies. In: Grau, et al, eds. (2009)

Ren, Y., Gröner, G., Lemcke, J., Rahmani, T., Friesen, A., et al.: Validating process refinement with ontologies. In: Kendall, E.F., Pan, J.Z., Sabbouh, M., Stojanovic, L., Zhao, Y. (eds.) 5th International Workshop on Semantic Web Enabled Software Engineering (SWESE), CEUR Workshop Proceedings. vol. 524, pp. 1–15. CEUR-WS.org (2009b) ISSN 1613-0073, http://ceur-ws.org/Vol-524/swese2009_1.pdf

Ren, Y., Pan, J.Z., Zhao, Y.: Soundness preserving approximation for tbox reasoning in r. In: Grau, et al (eds.) (2009)

Vanhatalo, J., Völzer, H., Leymann, F.: Faster and more focused control-flow analysis for business process models through sese decomposition. In: Krämer, B.J., Lin, K.-J., Narasimhan, P. (eds.) ICSOC 2007. LNCS, vol. 4749, pp. 43–55. Springer, Heidelberg (2007)

Wyner, G.M., Lee, J.: Defining specialization for process models. In: Organizing Business Knowledge: The MIT Process Handbook, ch. 5, pp. 131–174. MIT Press, Cambridge (2003), http://ccs.mit.edu/papers/pdf/wp216.pdf

Author Index

GPSR Compliance

The European Union's (EU) General Product Safety Regulation (GPSR) is a set of rules that requires consumer products to be safe and our obligations to ensure this.

If you have any concerns about our products, you can contact us on ProductSafety@springernature.com

In case Publisher is established outside the EU, the EU authorized representative is:

Springer Nature Customer Service Center GmbH
Europaplatz 3
69115 Heidelberg, Germany

Batch number: 09490872

Printed by Printforce, the Netherlands